TSO-DSO Interactions and Ancillary Services in Electricity Transmission and Distribution Networks

Gianluigi Migliavacca
Editor

TSO-DSO Interactions and Ancillary Services in Electricity Transmission and Distribution Networks

Modeling, Analysis and Case-Studies

Springer

Editor
Gianluigi Migliavacca
Energy System Development Department
Ricerca sul Sistema Energetico
Milano, Italy

ISBN 978-3-030-29202-7 ISBN 978-3-030-29203-4 (eBook)
https://doi.org/10.1007/978-3-030-29203-4

This Springer imprint is published by the registered company Springer Nature Switzerland AG
The registered company address is: Gewerbestrasse 11, 6330 Cham, Switzerland

Preface

During 3.5 years, I had the honour to coordinate the Horizon2020 project SmartNet (http://smartnet-project.eu), whose aim was to analyse different TSO-DSO coordination schemes and real-time market architectures. This theme is now of topical interest because of the fact that distribution networks are changing rapidly from "passive" (populated only by loads with no flexibility in their behaviour) to "active" (enclosing several kinds of distributed resources and storage). These distributed resources could become suitable to contribute to the provision of services to the system, but this implies a tighter coordination between the network operators (transmission and distribution system operators, TSO and DSO).

So, the "core" result of the SmartNet project was a comparison of different TSO-DSO interaction schemes and different real-time market architectures in order to find out which would deliver the best compromise between costs and benefits for the system. In order to reach such target, an ad hoc simulation platform was developed, and three physical pilots were deployed, representing each the practical implementation of a different TSO-DSO coordination scheme and testing a specific technological and ICT solution.

So, it came to me the idea it could be interesting for a wide public to acquire such kind of experience. The topic of TSO-DSO interaction for the provision of system services, yet being nowadays the subject of an important discussion and of a significant regulatory evolution in Europe, has been so far the object of very scanty publications. To my knowledge, a comprehensive book providing all foundations on this subject is still completely lacking. However, the basic idea here is not just to list a series of results of the SmartNet project but, rather, *to provide the reader with a theoretically solid background, so as to allow this book to be easily used by university-level courses interested to deal in detail with architectures of real-time services markets. The style followed is kept simple, and no preliminary background (apart from a basic knowledge of the electric system and of what electricity markets are) is given for granted.* First, the context is set, along with the problems affecting the electricity system. Then, different "benchmark" TSO-DSO coordination schemes are introduced. Subsequently, a background is provided on the mathematical modelling concepts most useful for investigating this kind of problems: a basis on ancillary services markets, network modelling (transmission and distribution), aggregation and disaggregation. A chapter on ICT requirements is provided too. Then, simulation scenarios are introduced, as well as the physical pilots which were developed in the SmartNet project. The last section of the book is dedicated to regulatory conclusions. A few reflections on parallelisms and differences between the two commodities, electricity and natural gas, are also reported.

Particular care is taken everywhere not to clutter the different chapters with too many formulas: for didactics' reasons, preference is given to illustrate basic concepts. However, a thorough bibliography is provided for the reader interested to get deeper on single topics: not only the relevant SmartNet deliverables but a lot of other scientific publications, regulatory documents, etc.

I hope this book will be useful in order to fill the already mentioned gap of didactic texts tackling in a simple but rigorous way the topical theme of TSO-DSO interaction for the provision of real-time ancillary services. In fact, tackling this theme needs a multidisciplinary approach involving knowledge of networks, market architectures, ICT and regulator aspects. The ambition of the present book is, thus, to provide a simple but complete introduction.

Milano, Italy Gianluigi Migliavacca

Contents

Chapter 1
Introduction

Gianluigi Migliavacca

The electricity world is rapidly changing. Renewable Energy Sources, often characterized by an important variability in their generation pattern, are growing significantly, also as an effect of the incentivization policies put in place during the last years.

Another important aspect is that a large amount of such generation is being deployed in distribution networks (e.g. photovoltaic power plants). Such networks were thought for distributing to the loads the energy produced by large generators located in transmission networks. The fact an important amount of generation is being deployed in distribution prompts them to become active grids, with power often remounting towards distribution, which constitutes a problem since this is not the exercise modality distribution grids were planned for.

However, this is not the only big change interesting distribution. Loads connected to distribution grids are not any longer thought of being sheer "energy-consuming" entities but are being more and more transformed into something which could provide flexibility services to the grid. Interruptibility of big industrial loads is already regulated in most European countries and special tariffs are applied to those loads which accept the risk to be interrupted within the terms specified by the contract. It is a sort of common understanding that the rated power of the loads which are allowed to behave in a flexible way and provide grids services will gradually decrease up to concern households and, in the future, local storage, electrical vehicles, etc. In this way, while, due to renewable energy sources, generation, will no longer be completely firm and adjustable according to market needs ("dispatchable" as it is said in the jargon), loads will become more and more dispatchable.

G. Migliavacca (✉)
Energy System Development Department, Ricerca sul Sistema Energetico, Milano, Italy
e-mail: Gianluigi.Migliavacca@rse-web.it

G. Migliavacca (ed.), *TSO-DSO Interactions and Ancillary Services in Electricity Transmission and Distribution Networks*,
https://doi.org/10.1007/978-3-030-29203-4_1

In any case a large amount of small-size generation and flexible load is expected to be deployed in the distribution grids. This will counteract the general tendency to decommission big generators connected to transmission networks, partially due to the fact they are no longer in the merit order (this happens especially for fossil fuel-based generation) or no longer complying with the demanding prescriptions on pollutants and CO_2 emissions. As such big power plants are traditionally those which provide resources for system balancing and congestion management in real time, that means small power plants and flexible load in distribution (in one word: distributed resources) will gain an increasing role in this domain: why not thinking of making such resources, once duly aggregated, become able to participate in real-time services markets (balancing, congestion management)?

However, that creates a coordination problem between transmission system operators (TSOs) and distribution system operators (DSOs): real-time markets (e.g. balancing markets) are usually managed by TSOs by getting access to resources connected to transmission networks, which are under their direct control. If a consistent amount of bids for real-time markets will come from subjects connected to distribution networks (yet through an intermediate aggregation process), the local DSO where resources are located should in some way become part of the process.

First of all, it is needed that the real-time status of the distribution networks becomes monitored. This is presently not the case, because the policy followed so far from DSOs was to provide for an over-dimensioning of their grids so as to be able to cope with the most demanding loads scenario ("fit-and-forget approach"). In a situation in which distribution networks are becoming active, this policy becomes more and more complicated to implement. Additionally, projects for building new lines encounter an increasing opposition from public opinion, on one side because of not being the most economically convenient solution (easily demonstrated by carrying out a cost-benefit analysis) and on the other because of the increasing sensibility of the public opinion to landscape issues and visual constraints. However, putting in place an extensive monitoring system is made complicated by the fragmentation of the DSO world: very rarely a DSO is covering the overwhelming majority of the territory in a country. Often, there are different sizes among DSOs and a topological hierarchy among them (primary DSOs directly connected to the TSO network, others only connected to the network of the primary DSO, etc.).

This last aspect is also the one that makes it more complicated to think of a new active role for the DSO in real-time markets. Many and variegated are the TSO-DSO architectures which can be implemented in order to do so. Whereas concerning system balancing, the current regulation is oriented to keep it firmly in the hands of the national TSOs, for what concerns congestion management, the new orientation of the EU regulation is that this role could be passed to local DSOs, which should become able to organize local congestion management markets for their own network. In Art. 32 of the *Proposal for a Directive of the European Parliament and of the Council on common rules for the internal market in electricity (recast)* part of the *Clean Energy for All Europeans* package, we can read: "Member States shall provide the necessary regulatory framework to allow and incentivise distribution system operators to procure flexibility services, including congestion management in their service area, in order to improve efficiencies in the operation and development of the distribution system". Such statement, for all reasons listed above, constitutes a big change with respect to the schemes implemented so far and constitutes a challenge for the next years both on the ground of its implementation and on the consequences it could have in respect to the balancing market still managed by the national TSO. First of all, local markets are by their own nature more subject to issues tied with scarcity of liquidity or incumbency of local operators. Both such problems can turn local markets into a failure or in any case into something well far away from markets behaving according to the price revelation principle, which constitutes the main reason competition and markets were introduced years ago with the so-called third package. Additionally, another big issue is represented by the coordination need between the congestion management managed by the local DSO and the system-wide balancing market managed by the TSO. First of all, it is important that the actions carried out in these two markets are somewhat coordinated or at least not in contrast, as well pointed out in the CEER Position Paper on the Future DSO and TSO Relationship – Ref. C16-DS-26-04 – 21.09.2016: "Some actions can have a negative cross-network effect. For instance, TSO use of distributed resources for balancing purposes has the potential to exacerbate DSO constraints. Equally, whilst DSO use of innovative solutions, such as active network management, can deliver benefits to customers, if not managed properly they may in some cases counteract actions taken by the TSO".

Last but not least, between the different markets, a "common marketplace" (see ENTSO-E working paper on Distributed Flexibility and the value of TSO/DSO cooperation) should be created, allowing to share bids databases by allowing them to be used by more than one market. This would avoid duplicating bids but also prevent different market operators from carrying out double activations.

It is also important that the services markets are able to fully account for the peculiarities and constraints of the bidding resources by offering complex products enabling such resources to compete on a level playing field ground with conventional generation, traditional providers of ancillary services and characterized by a higher level of flexibility and less operative constraints. That could bring to the necessity to adopt more complicated market architectures (e.g. allowing complex or mutual exclusive bids) which might considerably increase the complexity of the mathematical modelling and the computational time required for solving it.

Conceiving a local congestion market operator role for the DSO is not the only solution which can be put in place. The different TSO-DSO architectures are characterized each by a different role for the DSO in one single or two separated markets allocating resources for congestion management and system balancing. The architecture which reflects best the current status quo is the centralized one, in which the TSO manages a market in which both transmission and distribution resources may bid, while such market doesn't consider constraints and real-time status of the distribution grid. In this architecture, the only role reserved to the DSO is the one to provide a pre-qualification of the distribution resources, allowing them (or not) to bid in the services market. Of course, lacking a real-time monitoring of the distribution sector, this pre-qualification could be only done in a static way so as to ensure not to incur in congestion whatsoever is the current situation. Such scheme can't be very efficient but has the advantage to be very simple, less demanding from the computational point of view and simpler in the mutual coordination needs of TSO and DSO.

On the opposite side, we can conceive a scheme where TSO and DSO are managing together a market for congestion and balancing (possibly by appointing a separate subject charged of the role of market operator). This solution is for sure more efficient but it requires to include into the clearing algorithm a full management of line congestion for transmission and distribution. So, it is much more computationally demanding.

Analysing different TSO-DSO coordination schemes and real-time market architectures is the main goal of the Horizon 2020 project SmartNet (http://smartnet-project.eu). This project, started in January 2016 and concluded in June 2018, aims at comparing different TSO-DSO interaction schemes and different real-time market architectures in order to find out which would deliver the best compromise between costs and benefits for the system and answer to the following questions:

- Which ancillary services could be provided from entities located in distribution networks?
- Which optimized modalities could be used for managing the network at the TSO-DSO interface?
- How should the architectures of dispatching services markets be consequently revised?
- What ICT on distribution-transmission border guarantees observability and control?
- Which could be the regulatory implications?

In order to reach such target, an ad hoc simulation platform was developed, able to model in detail physical network, ancillary services markets and aggregation/disaggregation processes of the flexibility bids. Three national cases were analysed (Italy, Denmark, Spain) on scenarios at the target year 2030. Subsequently, this simulation platform was also implemented in a full replica lab, in order to test the performance of real controller devices. An ad hoc cost-benefit analysis was developed in order to compare the economic efficiency of the different coordination schemes and quantitative results were delivered. Such quantitative results fed a regulatory analysis which identified barriers and enablers to implement the TSO-DSO coordination schemes studied in the project and formulated regulatory guidelines.

Furthermore, SmartNet featured three physical pilots representing each the practical implementation of a different TSO-DSO coordination scheme and testing a specific technological and ICT solution:

- Technical feasibility of key communication processes enabling small run-of-the-river hydro generators located in a remote Alpine region in Südtirol to participate to frequency and voltage regulation by exchanging signals with the Italian TSO Terna (Italian pilot)

- Capability of flexible demand to provide ancillary services for the system:
  - Thermal inertia of indoor swimming pools (Danish pilot)
  - Distributed storage of base stations for telecommunication (Spanish pilot)

The SmartNet consortium, under technical and administrative management by RSE, included 22 partners from 9 European countries, among which are TSOs (Energinet.dk, Terna), DSO (Endesa, SydEnergi, Edyna), manufacturers (Selta, Siemens) and telecommunication companies (Vodafone).

The European research project SmartNet constitutes the background for the material presented in the present book. However, this book, yet illustrating material and investigations derived from the SmartNet project activities, has a different aim and scope. As a matter of fact, the idea underlying the book is to provide the reader with a theoretically solid background allowing it to be easily used by university-level courses interested to deal in detail with architectures of real-time services markets. The style followed is kept simple and no preliminary background (apart a basis knowledge of the electric system and of electricity markets) is given for granted.

The subsequent chapters have the following structure:

- Chapter 2 – *TSO-DSO Interaction and Acquisition of Ancillary Services from Distribution* – it explains the characteristic of the coordination schemes taken as reference for the investigation and the cost-benefit analysis.
- Chapter 3 – *Modeling of Complex Systems Including Transmission, Distribution, Aggregation, Ancillary Services Markets* – it provides the background for the mathematical modelling adopted to describe the system.
- Chapter 4 – *ICT Requirements in a Smart Grid Environment* – it provides a reference methodology for the investigation of ICT requirements to implement different TSO-DSO coordination schemes.
- Chapter 5 – *Scenario Analysis* – it describes the scenarios at the 2030 target year adopted for the simulation of the three national cases: Italy, Denmark and Spain.
- Chapter 6 – *Technologies and Protocols: The Experience of the Three SmartNet Pilots* – it presents the main returns of experience collected by the three SmartNet national pilots.
- Chapter 7 – *Regulatory Frameworks for Enabling Distributed Energy Resource Participation in Smart Grids* – it presents regulatory considerations in relationship with the possible implementation of the studied coordination schemes, also in relationship with current regulatory trends in Europe.
- Chapter 8 – *Conclusions* – Beyond providing conclusive remarks, it also casts a parallel to the regulatory evolution in the sector of natural gas, commodity which, yet retaining important differences with respect to electricity, also has a lot of characteristics in common. The parallel between electricity and gas gives the opportunity for a few interesting considerations.

Chapter 2
TSO-DSO Interaction and Acquisition of Ancillary Services from Distribution

Helena Gerard, Enrique Rivero, and Janka Vanschoenwinkel

2.1 Introduction

The electricity grid is a large network[1] ensuring power delivery through the connection of numerous individual electricity producers and consumers. For it to work in a stable, efficient, and reliable way, some important requirements need to be fulfilled. First of all, supply and demand (generation and load) of electricity must be balanced,[2] every second of the day (in real time), because (today) electricity cannot be stored in an economically viable way. Second, generation and load need to be adjusted to manage power flows within physical grid constraints.

Because of the physical grid reality, an optimal operation strategy is needed. The actors responsible for the planning, development, and stable operation of the transmission and distribution grids are, respectively, the transmission and distribution system operators (TSO and DSO). To fulfill their responsibilities, system operators (SOs) can make use of system services. Services used by the TSO are commonly known as ancillary services (AS); services used by the DSO are called local system services. These services can be provided by resources connected to the electricity grid and they can be called upon when needed. There is a difference in the resources that the TSO and the DSO use. The TSO is typically using resources connected to the transmission grid, which consist mainly of large conventional power plants (e.g., nuclear, gas, and coal). In most deregulated European power systems, the TSO is currently responsible for procuring services to also maintain the reliable operation

[1] Consisting of a distribution and a transmission grid.

[2] System balancing implies among others that the TSO has to correct for instant deviations between injection and off-take of electricity. This is indispensable as such imbalances could lead to a drop in frequency or voltage.

H. Gerard (✉) · E. Rivero · J. Vanschoenwinkel
EnergyVille/VITO, Genk, Belgium
e-mail: Helena.gerard@vito.be

G. Migliavacca (ed.), *TSO-DSO Interactions and Ancillary Services in Electricity Transmission and Distribution Networks*,
https://doi.org/10.1007/978-3-030-29203-4_2

of the interconnected system (this includes AS).[3] The DSO on the other hand mainly relied on grid reinforcements through additional cable connections to avoid operational issues [17]. Unlike the TSO, the DSO is currently not yet relying on system services to be used for local balancing, voltage control, or congestion management.

Today, the provision of AS is changing. Thanks to advancement in technology (monitoring and control, smart meters, etc.), harvesting flexibility from distributed resources has become possible. At the same time, the energy market is undergoing important changes because of the continuous development of the European internal electricity market and the increase of distributed energy resources (DER). DER encompass a large range of flexibility resources such as distributed generation (DG), storage resources, and demand-side management (DSM). As a result, the provision of AS is becoming more diverse and is not only provided by transmission-connected generation.

This increased availability of DER could bring along opportunities for managing the power system[4] as the flexibility from these resources could be used to provide system services to both TSO and DSO. The variable generation located in the distribution grid could be combined, for instance, with local storage and active demand management to provide local services for the distribution grid (voltage control and congestion management). For the TSO DER could be used to provide services for the entire system through the connection point between the distribution and the transmission grid (frequency control, voltage control, or congestion management) [6, 10, 15, 22].

If SOs want to make optimal use of DER, they will have to establish a stronger coordination (up to real time) among them and also between other actors that are involved in the provision of AS [5, 10, 21]. This coordination is needed to contribute to a more efficient and cost-effective operation of the relevant grids involved, for instance, by ensuring that actions taken by one SO would not contradict actions taken by another SO [3, 4, 6, 7, 10, 14, 16, 17, 20, 21].

Current levels of TSO-DSO coordination could be hindering opportunities to exploit DER participation in AS. At European level, the amount of contracted volume of DER for system purposes is still limited as is shown in the SmartNet project[5] that evaluated a sample of European countries (Austria, Belgium, Denmark, Finland, Italy, Norway, and Spain). Even though there were clear differences across

[3] From a simplified view, the TSO provides these services to respond to a few tasks (balance generation and load, maintain voltages, control generation to avoid overloading of grid assets, restart the system (after a collapse)).

[4] Challenges related to DER are, for instance, related to the fact that they are much more dispersed (smaller volumes) and often weather dependent (wind, solar). The latter makes them more volatile and less predictable which could lead to system imbalances and increase network operators' need for control services closer to delivery.

[5] The SmartNet project arises from the need to find answers and propose new practical solutions to the increasing integration of renewable energy sources in the existing electricity transmission network. This project has received funding from the European Union's Horizon 2020 research and innovation program under grant agreement no 691405. http://smartnet-project.eu/.

countries, in most countries the TSO was contracting directly resources connected to the distribution grid, except for Italy where DER were not yet allowed to provide AS (end of 2016). Currently, in none of the countries, the DSO was procuring flexibility for its own needs. On the one hand, this is explained by the fact that the DSO is not always allowed to contract flexibility for operational purposes. Yet, on the other hand, in countries where such contracting was allowed, there was no financial incentive for DSOs to do so as costs for the procurement of flexibility are not always recognized as eligible to be reimbursed. This leaves reinforcing the grid instead of using flexibility as the most attractive option for DSOs to solve local congestion.

To discuss procurement of AS from DER, five coordination schemes (CSs) are proposed. These CSs focus on the interaction between TSOs and DSOs for the procurement and activation of AS. The chapter continues as follows: Section 2.2 elaborates on the role of SOs and presents the five models for coordination between TSO and DSO. Section 2.3 assesses the benefits and attention points for each of the CSs. Section 2.4 examines the feasibility of each CS in relation to the current European and national context. Section 2.5 concludes.

2.2 The Role of SOs in the CSs

A CS can be understood as the relation between a TSO, DSO, and other market players, defining the roles and responsibilities of each actor, when procuring and using system services provided by resources connected to the distribution grid [12, 18]. As such, the schemes determine the operational processes and information exchanges between SOs and other market players when procuring flexibility-based services.

As defined by Gerard et al. [12] and Rivero et al. [19], a role is an intended behavior of a specific market party, with certain responsibilities, which is unique and cannot be shared. There are different roles related to the different steps (prequalification, procurement, activation, and settlement) needed for the planning, provision, and acquisition of services by SOs. The difference between CSs is mainly observed in the activity of the procurement of AS. This is because dependent on the CS, stakeholders might take up different roles in the process of the procurement of services, which also has a significant impact on the interactions between the different stakeholders. Roles related to the processes of prequalification, activation, and settlement of flexible resources are undoubtedly assigned to a specific market party. As such, these processes are rather similar across CSs. The main influence of the CS is therefore situated in the procurement phase of the AS or local system services [11].

The key roles that are involved when procuring AS are the role of the reserve allocator (RA), the buyer, the seller, the market operator (MO), and the aggregator. The RA is responsible for determining the amount of flexibility-based services (e.g., reserves) that need to be procured; the buyer is the actor acquiring the flexibility-based services in a market setting; the seller is the actor providing these services; the

MO is responsible for setting up the market platform and operating and clearing the market; and the aggregator is collecting DER flexibility from different suppliers for its offering in a market setting.

Table 2.1 provides an overview of the most relevant roles that have to be considered in the context of the prequalification, procurement, activation, and settlement of AS. For each role, it is indicated which market party could take up this role.

In what follows, we present the five CSs, ranging from a centralized AS market model to an integrated flexibility market model (See Fig. 2.1).

Centralized AS Market Model In this CS, there is one centralized, common market for AS for both resources connected at transmission and at distribution level. This market is operated by the TSO, independent of the fact whether the resources are connected at the transmission or at the distribution level. The TSO determines the technical needs to operate the system in real time and communicates the required amount to the market. In doing so, the TSO does not actively take DSO grid constraints (such as capacity limits) into account.

As a result, the role of the DSO is limited: the TSO contracts flexibility from DER directly from the DSO grid; and the DSO is not involved in the procurement and activation process of AS by the TSO. The DSO does not procure local flexibilities in real time or near to real time to solve local grid issues.

In order to respect the DSO grid constraints, the TSO could install a separate process of system prequalification to ensure that the activation of resources from the

Table 2.1 Roles for prequalification, procurement, activation, and settlement of AS

	Role	Explanation	Adopted by
Grid operation	System operator (SO)	Operates and manages the physical system in question	TSO; DSO
	System balance responsible (SBR)	Ensures the balance of the grid and reduces deviations for a system or certain area by the activation of reserves	TSO; DSO
	Data manager (DM)	Handles grid data (incl. formatting, storage, and provision), separately for each network level	TSO; DSO; IMO
Pre-qualification	Flexibility feasibility checker (FFC)	Responsible for assessing potential impact at distribution grid level (system prequalification) caused by the provision of flexibility-based services from a DER unit requesting participation to the AS flexibility market (central or local)	DSO
Procurement	Reserve allocator (RA)	Determines the amount of flexibility-based services (e.g., reserves) to be procured	TSO; DSO
	Buyer	Acquirer of flexibility-based services in a market setting	TSO; DSO; CMP
	Seller	Provider of flexibility-based services in a market setting	TSO; DSO; CMP;
	Market operator (MO)	Responsible for setting up the market platform and operating the market	TSO; DSO; IMO

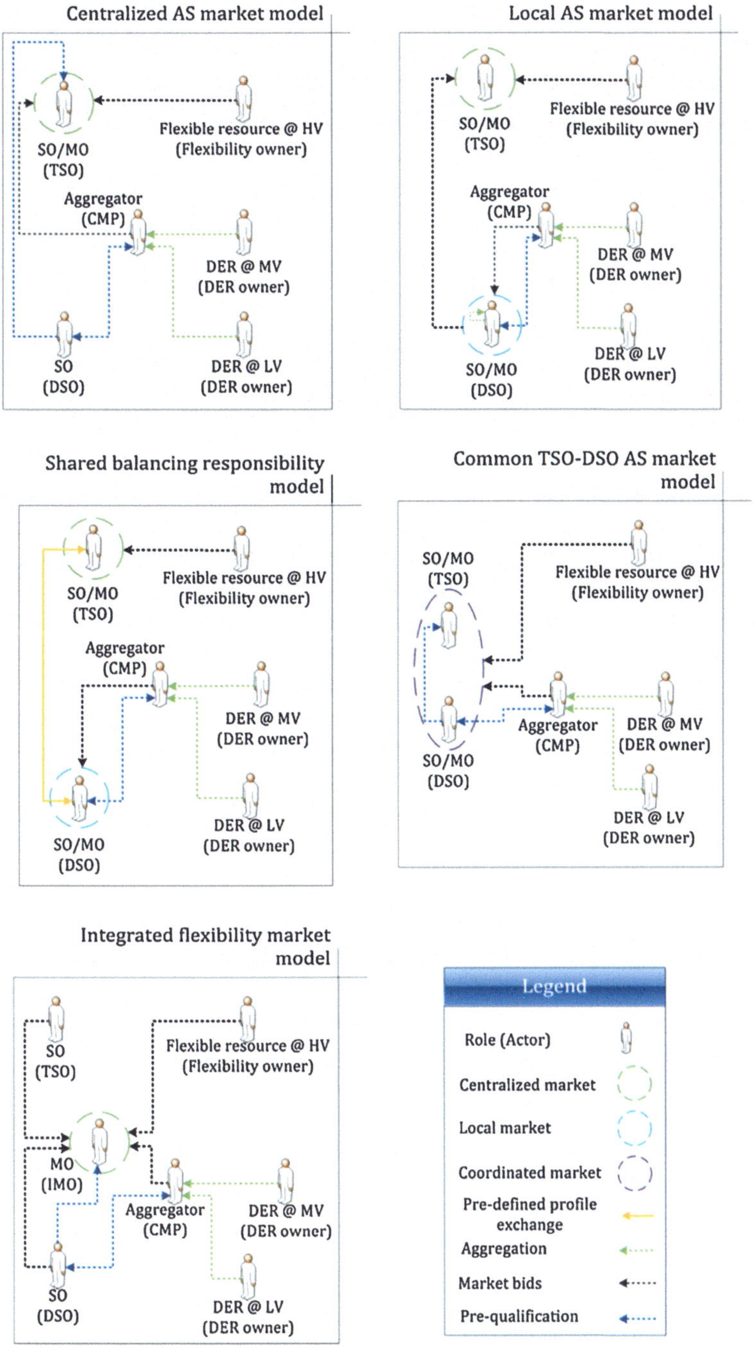

Fig. 2.1 CSs: high-level view of roles, market architecture, and stakeholder interactions

distribution grid by the TSO does not cause additional constraints in the distribution grid (e.g., congestion). In that case, the DSO is minorly involved in the process of system prequalification as it needs to provide the necessary data to the TSO.

Local AS Market Model In this CS, a separate local market for system services is implemented, operated by the DSO. The clearing of the local market happens first and the DSO contracts the necessary flexible resources to be used for local congestion management. The TSO is operating a central market for AS where resources connected to the transmission grid can participate, next to aggregated resources from the distribution grid that were not selected in the local market

Sellers of flexibilities will send aggregated bids of flexibility to either the local market or the AS market, depending on whether they are connected to the distribution or transmission grid, respectively. When the market operator of the local market (DSO) clears his market, he respects the local constraints and he keeps the selected bids for his own local use. The key difference with the centralized AS market model is therefore that the DSO always has priority to use local resources for local congestion management and distribution network constraints are respected as the local market is operated by the DSO itself. As such, the local market shifts priorities toward the DSO to first solve potential local congestion. Only flexibility not needed/procured at the local market is sent to the central TSO market, which is then cleared as well. The role of the TSO therefore remains limited to the operation of its own AS market. Nevertheless, note that in case the DSO activates contracted flexibilities to solve local congestion, in order to negatively impact the overall system balance, the DSO has to counter activate another bid in the opposite direction. Consequently, it is important that the TSO is informed in order to avoid the possible counter activation in case this is not necessary from a system perspective (e.g., the activation of the bid to solve the local congestion was at the same time restoring the existing system imbalance).

Shared Balancing Responsibility Model Like the local AS market model, the shared balancing responsibility model also has an AS market for resources connected to the transmission grid (operated by the TSO) and a local market for resources connected to the distribution grid (operated by the DSO). However, compared with the local AS market model, the difference is that resources from the distribution grid cannot be offered or transferred to the central AS market. The role of the DSO is therefore new in the sense that it is responsible for balancing the distribution grid according to a schedule defined upfront between the TSO and the DSO.[6] As a result, the DSO will procure local flexible resources to solve both

[6] Such a pre-defined schedule is necessary as TSO and DSO grids do not work independently from each other (they are connected through interconnection points). Through these interconnection points, there is a flow of energy. This flow can (for instance) be predicted in advance based on BRP nominations in the day-ahead market. A nomination tells the TSO the planned generation and/or consumption of the BRP. In the shared balancing responsibility model, these nominations could be used to develop each day a pre-defined schedule that determines the flow through the interconnection points.

local congestion and local balancing. This implies that the shared balancing responsibility is the only scheme where the TSO has no direct access to resources connected at the distribution grid. The flexibility offered by the distribution grid is only reserved for the DSO.

Compared to the local AS market model, the first procurement step in the shared balancing responsibility model is different as there should first be a mutual agreement between the SOs (DSO and TSO) with regard to the exchange profile. All the other steps are similar to the local AS market model procurement step. Only the market clearing is different as both markets are cleared simultaneously while fulfilling both SO's requirements.

With regard to the pre-defined schedule, it should be noted that it could be specified in two different ways. On the one hand, one can determine the schedule at the level of the entire DSO area, meaning that the nominations or flows through each individual TSO-DSO point do not need to be known. This method implies that only one aggregated schedule is determined (the net exchange) and that only the outcome of energy-only markets is needed. The method is easier to calculate and requires few interactions between TSO and DSO. This comes, however, at the cost of being unable to account for real-time or near-to-real-time constraints at the TSO-DSO interconnection. On the other hand, TSOs and DSOs could determine a schedule very specifically for each TSO-DSO interconnection point. This implies that historical forecasts at each TSO-DSO interconnection point are needed, together with congestion constraints in the transmission and distribution grid. Even though it is very precise, this method is harder to calculate since it requires close cooperation between SOs (such as sharing of relevant and accurate data), which might be challenging in a short timeframe.

Common TSO-DSO AS Market Model In the previous shared balancing responsibility model, the TSO did not have access to flexible resources connected to the distribution grid. Nevertheless, there are arguments that purchases of flexibility services by the TSO should be possible, either channeled through DSOs or coordinated together with DSOs by implementing common auctions [1].

This common TSO-DSO AS market model therefore proposes a common market for flexible resources connected to both the transmission and the distribution grid. The TSO and the DSO operate the market jointly in a way that the outcome for the system as a whole is optimized. In the end, the TSO contracts AS services from both transmission and distribution, and the DSO uses flexible resources from its own grid in cooperation and in interaction with the TSO. As a result, this CS has the potential to minimize total procurement costs. Unlike the other schemes, this implies that in this scheme there is no upfront priority for the TSO or the DSO.

How this could work in practice depends on how the concept of the common market is implemented (e.g., one single platform or multiple decentralized platforms that are connected with each other).

Integrated Flexibility Market Model As with the Common TSO-DSO market model, this integrated flexibility market model provides a common market for flexible resources connected to the transmission and distribution grid. However, unlike any other CS, this scheme introduces the participation of both regulated (SOs) and deregulated (commercial market parties) parties to procure flexibility in a common market. All market players (SOs and other commercial market players (CMPs)) that need flexible resources or that have flexibility sources to offer communicate their needs or bids to the market operator.[7]

The difference between this scheme and the other schemes is therefore that it allows direct competition between regulated and nonregulated players under the same conditions. For this to work, the presence of an independent market operator is required to ensure neutrality. Compared to the other CSs, this implies a new entity that will have an influence in the management of data and the settlement of the market. If market neutrality is ensured, it will be the market forces that dictate how flexibility will be allocated: resources are allocated to the party with the highest willingness to pay.

Competition is therefore introduced at the demand side, and the SOs might not necessarily receive what they have asked for. However, this might not be a problem if settlement rules encourage market parties to support system operation. As a result, this CS is a more futuristic and theoretical approach toward the consideration of flexibility needs. The market is not offering services anymore, but resources without a specific objective (e.g., flexibility).

2.3 Assessment of CSs

Comparing the different CSs, the attentive reader will have noticed that going from the first to the last scheme, the role of the DSO in the procurement of AS increases. Table 2.2 gives an overview of this gradual expansion of the DSO role.

The choice for a specific CS impacts the TSO grid operation, the DSO grid operation, other market participants, and the functioning of the market. To more properly understand the benefits and disadvantages of the CSs, we present below a high-level qualitative assessment of the CS on selected performance criteria. Table 2.3 gives an overview.

2.3.1 Centralized AS Market Model

This CS is "simple" in the sense that there is only one market place, operated by the TSO. This unidirectional market operation makes market functioning relatively straightforward and market products are clearly known to all market participants.

[7]Note that system operators (TSO and DSO) could also decide to sell previously contracted flexibility.

Table 2.2 Summary of the key elements of the five CSs with respect to the role of the DSO, the organization of the market, and the allocation principle of flexibility from the distribution grid

Coordination scheme	Role of the DSO	Market organization (market operator)	Allocation principle of flexibility from the distribution grid
Centralized AS market model	Limited to possible process of prequalification	Common market (TSO)	Priority for the TSO
Local AS market model	Organization of local market Buyer of flexibility for local congestion management Aggregation of resources to central market	Central market (TSO) Local market (DSO)	Priority for the DSO
Shared balancing responsibility model	Organization of local market Buyer of flexibility for local congestion management and balancing	Central market (TSO) Local market (DSO)	Exclusive use for the DSO
Common TSO-DSO AS market model	Organization of flexibility market in cooperation with TSO Buyer of flexibility for local congestion management	Common market (TSO and DSO) Central market (TSO) Local market (DSO)	Minimization of total costs of TSO and DSO
Integrated flexibility market model	Buyer of flexibility for local congestion management	Common market (independent market operator)	Highest willingness to pay

Source: Gerard et al. [12]

Table 2.3 Overview of benefits and risks of the five CSs

Domain	Main benefits and risks	Coordination scheme				
		Centralized AS market model	Local AS market model	Shared balancing responsibility model	Common TSO-DSO market model	Integrated flexibility market model
Interaction between system operators	Need for increased communication between system operators	Low	Medium	Medium	High	Medium
	Need for sharing of data between system operators	Low	Medium	Low	High	Medium
Grid operation	Risk of violation of distribution grid constraints	Medium	Low	Low	Low	Low
	Possibility to have access to resources from the distribution grid by the TSO	High	Medium	Low	High	High
Market functioning	Operational costs of organization of the market	Low	High	Medium	High	High
	Risk of illiquidity in the market	Low	High	High	Low	Low

Source: Gerard et al. [12]

The main advantages of this organization are that only little additional communication infrastructure is required between different SOs and that costs for market operation are low. This is mainly because this CS is closely related to the current market functioning and because the structure therefore does not have to go through fundamental changes. Moreover, flexibility service providers can use resources from different DSO areas with one common bid and the simple market allows for standardized processes.

However, this market structure also has some drawbacks; i.e., there is only a limited involvement of the DSO in this CS. This implies that distribution grid constraints are only taken into account to a limited extent, for example, in case there is a process of system prequalification by the DSO installed. DSOs do not use flexibilities to solve local grid constraints in the same timeframe as the TSO.

2.3.2 Local AS Market Model

The local AS market model ensures through a local market operated by the DSO that DSO grid constraints are taken into account and that the DSO can activate flexibility resources in order to solve local constraints. In addition, if well-designed, such smaller local markets might create better conditions for smaller scaled DER (e.g., lower entry barriers).

However, such benefits highly depend on the number and the size of the market players in the market. In case there are multiple small distribution grids, each having a separate local market, there is a risk of having highly fragmented markets, which in turn could limit their liquidity[8] because each DSO would act individually as an aggregator for the bids it receives on its own separate market. As a result, optimizing bidding strategies becomes more difficult as different flexibility resources are now spread out over different separate local markets. This limits the amount of possible combinations of flexible resources, aggregated into one bid, as resources now belong to different DSO areas. The price for aggregated bids for the TSO might therefore be less optimal (more expensive) than prices determined for each individual DSO area. Moreover, market fragmentation might lead to different market products in multiple local markets. This can be both positive (in the sense that tailor-made products to the needs of specific market players are developed) and negative in the sense that aggregation becomes more complex. There might therefore be a need for harmonization regarding such aggregated bids.

For communication, consequences of having multiple local markets are that communication between all such markets and the central market becomes more

[8] Liquidity refers to the speed and the ease with which investors can realize the cash value of an investment. Illiquid assets, for example, real estate, can be hard to sell quickly, and a quick sale may require a substantial discount from the price at which it could be sold in an unrushed situation [2].

extensive and ICT infrastructure harder to implement. In the latter case it should therefore be considered whether an additional layer of aggregation over multiple separate local markets (e.g., a shared platform) could be more optimal. This could reduce costs and make the market more efficient.

2.3.3 Shared Balancing Responsibility Model

Within this CS, part of the balancing responsibility is now transferred to the DSO according to a pre-defined schedule. The flexible resources connected to the distribution grid can be used exclusively by the DSO.

However, the higher amount of flexibility that needs to be procured (because the DSO needs resources for both local congestion management and local balancing) might lead to liquidity risks in case there are multiple small distribution grids, leading to similar issues as within the local AS market model (e.g., higher procurement costs, less possibilities for aggregation into one common bid, etc.). In the most pessimistic scenario, all this might lead to a DSO not being able to contract sufficient flexibilities (especially in smaller distribution grids). If this occurs, the DSO might need to take unwanted measures such as curtailment or load shedding. In addition, the DSO is obliged to set up its own system for determining and billing imbalance penalties to BRPs. This system would have to be separate from the TSO system which might lead to additional operational costs for this CS. Moreover, because the TSO does not have access to resources at distribution grid, there are less options for the valorization of flexibility from the distribution grid.

2.3.4 Common TSO-DSO AS Market Model

In this CS, TSOs and DSOs look together for a combined solution that leads to cost minimization. This has as main advantage that grid costs are optimized. In addition, this CS could be a basis for further collaboration and/or integration between SOs. Dependent on the chosen variant, the market might be organized as a common market (centralized variant with a common market platform) or as a set of local markets dynamically connected to a central market (decentralized variant).

For the centralized variant, the main advantage compared to the centralized AS market model is that the operational costs of the market operation are shared between the DSO and the TSO, leading to an incentive to minimize total grid costs. Other benefits that are applicable to the centralized AS market model (such as simple processes and the possibility for standardized products) remain. However, to accomplish this, both the TSO and the DSO will need to share data with the common market. Data knowledge is important to establish a clear framework to manage SO interactions to allow for efficient whole system outcomes.

Comparing the centralized variant with the decentralized variant, the costs of the DSO (as market operator of the local market) are higher in the decentralized variant of the common TSO-DSO AS market model. In addition, like in the local AS market model, there might be less possibilities to aggregate several resources into one common bid (e.g., due to fragmentation). Also, with regard to the market functioning, communication requirements are lower in the centralized market, compared to in the decentralized market. Yet, there is still a need to share data with the common platform. This requires clear rules for security and privacy of data. However, the structure of the decentralized market allows local bids to be submitted directly to the local market which might be less complex than when bids are offered directly to the TSO.

2.3.5 *Integrated Flexibility Market Model*

In this CS, there is one common market place for all flexibility providers and customers (i.e., for all market participants). This leads to increased possibilities for BRPs to balance their portfolios, for SOs to resell unneeded procured resources or for SOs to get access to unneeded previously contracted resources from other market players in case of wrong estimations. As a result, the costs for procurement of flexibility might be lower due to the higher liquidity of the market. In addition, given the fact that there is one common market, operational costs for individual parties are lower as costs are shared over a larger number of market participants.

However, for this market to function properly, neutrality is important and is ensured by the independent market operator. The presence of an independent market operator requires both TSO and DSO to share their data (on, for instance, distribution grid constraints) with this independent market operator. For this, clear rules for data security and privacy need to be established and agreed upon. Finally, responsibilities of the independent market operator need to be clarified toward all the other market participants.

Nevertheless, the presence of both regulated and nonregulated parties in one common market might raise specific concerns. First, it will be more complex for the TSO to determine the amount of AS to be procured as the commercial market parties (CMPs) can also buy flexibility almost in real time to keep their positions balanced. This will of course be mainly an issue at the start of the integrated market. If several market sessions have taken place, the TSO has a good view on the volumes typically needed. In addition, the TSO might still procure reserves outside the common market which could be used as an additional security measure. However, AS procurement of the TSO outside the market might impact the liquidity of the market itself and could be a barrier for the development of the integrated market. Another concern might be that opening the AS markets for CMPs might hinder further development of intraday markets.

2.4 Feasibility of CSs

When implementing a CS, in addition to the current market arrangement, and the roles and interactions taking place (especially between the roles adopted by the TSO and DSO), the following three points should be taken into account: the current regulatory framework, the organization of DSOs, and European trends toward harmonization and integration of electricity markets. This is important because adaptations in existing regulation might be required in some countries before a specific CS can be implemented. Therefore, current initiatives such as the ones looking to harmonize and integrate electricity markets could facilitate the implementation of a CS.

2.4.1 Regulatory Framework

The CS that is closest to be implemented today or in the near future (2020), without many changes, is the centralized AS market model. As of today, this scheme is most compatible with the current regulation and market organization in Europe.

With regard to the feasibility of other CSs, it is observed that their implementation highly depends on the evolution of roles of SOs and how they fit into the regulation. Today, the TSO has the core responsibility of balancing the AC power system.[9] However, this responsibility might be shared (or assigned in part) to the DSO (see Sect. 2.3.3).

Except for the *centralized AS market model*, all CSs assume that the DSO contracts flexibility resources to solve local grid constraints. Today, this is not done by DSOs due to several regulatory barriers (e.g., cost recognition, etc.). However, the recast of the Integrated Electricity Market Directive (IEM) promotes an active procurement and use of flexibility by the DSO. Current regulation is in most countries not yet adapted to this potential and/or new responsibility of DSOs. For instance, in most countries surveyed by the SmartNet project, DSOs are not often allowed or financially stimulated to contract flexibility (see Sect. 2.1). Specifically:

- For the common TSO-DSO AS market model, incentive regulation for SOs has to be changed to encourage the common objective of cost minimization.
- For the local AS market model and the decentralized variant of the common TSO-DSO market model, the DSO needs to be allowed to aggregate resources on behalf of the TSO.

Regulation therefore has to be adapted in the sense that DER participation is facilitated, and DSOs should be adequately remunerated when they improve efficiencies in the operation and development of the distribution system.

[9] From a simplified view, the operator of the AC power system must balance generation and load and maintain voltage across the power system in normal and contingency conditions, he controls that the loading of grid assets stay within safety margins, and he restarts the system after a collapse.

Finally, a common feature across CSs is the recognition of respecting distribution grid constraints. Therefore, several modifications to regulatory framework need to be made in order to support the implementation of the proposed CSs. The regulatory impact is discussed further in the following chapters.

2.4.2 *Organization of Distribution SOs*

DSOs can be organized in different ways in different regions. They can be large or small, they can be responsible for different grid voltage levels, there can be multiple of them or just a few within one country, etc. This existing organization of national systems highly influences the feasibility of implementation of different CSs.

In Sect. 2.2 it was indicated that the presence of multiple small DSOs could lead to low liquidity in some CSs. As a result, cost of flexibility procurement might increase and there are limited economies of scale. In countries (such as Germany and Norway) where this is a potential issue, this is an important aspect to take into account. Illiquidity of the local market could also be an issue when the available flexible resources in a given distribution grid are limited. This is therefore an aspect to be considered in the shared balancing responsibility model, in which the DSO has to comply with a scheduled profile for its area, contracting resources for both local congestion management and local balancing.

Another potential problem that could take place when the DSO is not the market operator is the communication of distribution grid constraints to the market operator. Depending on the definition of the AS product, this process will require a varying level of automation. For instance, for products that require to be provided in a short time span (seconds), a high level of automation would be required. This is not a problem in the schemes where the DSO is responsible for the local market organization.

An option to overcome the potential issues linked to many small markets might be to integrate or coordinate them. An example of this point is that it might, for instance, be more cost-efficient for the DSO, given its size, to procure flexibility via a market managed by a larger DSO or an independent body.

2.4.3 *European Trends Toward Harmonization and Integration of Electricity Markets*

One important trend is the realization of the European internal electricity market. Currently, efforts toward the IEM focus on further harmonizing and liberalizing European electricity markets [13]. During the 1990s, most national electricity markets were still monopolized, and the European Union decided to open up markets. The most recent proposal for a directive on common rules for the internal market in electricity [8] recasts Directive 2009/72/EC.

As a result, several initiatives are ongoing to promote the harmonization and integration of the European internal energy market. The market coupling of day-ahead markets is close to being finalized and the coupling of intraday markets is ongoing. This evolution of the intraday market is in particular relevant for the *integrated flexibility market model*. In particular, the ongoing integration of intraday markets is relevant for the integrated flexibility market model as both market models allow the participation of regulated and nonregulated market players. The integrated flexibility market model and the intraday market could coexist as long as they complement each other, for example, by offering different services in different timeframes. When services and/or timeframes overlap, there might be a need to integrate both market models. In addition, the markets for AS are also subject to further integration and harmonization.

This harmonization and integration of national markets is important in the context of the different CSs. Member states should therefore, in their choice for national CSs, take these ongoing initiatives into account to ensure that there is not a too large variety in national CSs

Nevertheless, the choice of a particular CS is not only country dependent. Different CSs might exist at the level of the individual product for different AS [9, 15]. For some AS, certain CSs are irrelevant. For AS used for balancing (e.g., FRR (automatic and manual), RR) or congestion management, every CS is possible. However, for AS related to frequency containment reserve (FCR) or voltage control at the transmission level, some CSs could be excluded.

2.5 Conclusion

Power systems are moving from a centralized to a more distributed structure. To a large extent, this evolution is explained by increasing amounts of DER that are connected to the distribution grid. This new form of electricity generation brings both challenges and opportunities for SOs as they can use these new resources as flexibility for the distribution grid. This flexibility can support, i.e., frequency control, congestion management, or voltage control. However, before it can be used to its full potential, coordination between SOs (TSO and DSO) is important. In particular, more active distribution grid management is needed as the DSO will also be expected to support the TSO in balancing the power system.

This book chapter presents a framework for policy makers to determine what are options to organize the cooperation between SOs concerning the procurement of AS. This is important to be able to exploit available flexibility at distribution and to understand how regulation could be adapted. The most suitable CS, however, highly depends on the starting point of the power system (e.g., the current organization with its local characteristics) and its long-term objectives (the type of flexibility service, the current state of the grid, the share of RES installed, the existing market design, and the evolution of roles and responsibilities of SOs). Yet, each of the CSs can evolve to more advanced CSs. In this way, policy makers can gradually introduce measures to intensify the coordination between SOs. Currently, it seems that the centralized AS market model is the closest to most power systems. In the short term, the local AS market model and the shared balancing responsibility model are probably easier to implement from an operational perspective. The other schemes could be implemented to increase liquidity and to take full advantage of economies of scale.

In practice, a change from one CS to another is in principle a question of a change in roles, responsibilities, and market design. Nevertheless, it is in particular the role of the DSO that will need to adapt to the CSs. In this regard, attention must be paid to the fact that the choice for a specific CS is embedded in the ongoing processes of harmonization and integration of power systems across the European Union. Finally, different CSs with different levels of interaction between SOs and other market players will impact business processes, information exchanges, communication channels, and ICT infrastructure. Making the necessary changes requires a paradigm shift in system operation.

References

1. C. Batlle, M. Rivier, Redefining the new role and procedures of power network operators for an efficient exploitation of demand side response (Working paper IIT) (2012)
2. Z. Bodie, A. Kane, A.J. Marcus, *Investments*, The McGraw-Hill/Irwin series in finance, insurance and real estate, 9th edn. (McGraw-Hill/Irwin, New York, 2011)
3. CEDEC, EDSO SG, ENTSO-E, Eurelectric, GEODE, General guidelines for reinforcing the cooperation between TSOs and DSOs (Technical report) (2015)
4. CEER, Position paper on the future DSO and TSO relationship (2016)
5. CIEP/PBL, Reflections on coordination mechanisms for accommodating increasing amounts of wind and solar in the power market (2014)
6. R. D'hulst, J.M. Fernández, E. Rikos, D. Kolodziej, K. Heussen, D. Geibelk, A. Temiz, C. Caerts, Voltage and frequency control for future power systems: The ELECTRA IRP proposal. 2015 International Symposium on Smart Electric Distribution Systems and Technologies (EDST). Presented at the 2015 International Symposium on Smart Electric Distribution Systems and Technologies (EDST) (2015), pp. 245–250. https://doi.org/10.1109/SEDST.2015.7315215
7. C. Eid, L.A. Bollinger, B. Koirala, D. Scholten, E. Facchinetti, J. Lilliestam, R. Hakvoort, Market integration of local energy systems: Is local energy management compatible with European regulation for retail competition? Energy **114**, 913–922 (2016). https://doi.org/10/gb2fs3
8. European Commission, Proposal for a directive of the European Parliament and of the Council on common rules for the internal market in electricity (recast) (2016)
9. European Commission, Draft regulation establishing a guideline on electricity transmission system operation – the System Operation Guideline (2016)
10. Expert Group 3, Regulatory recommendations for the deployment of flexibility (EG3 report). Smart Grid Task Force (2015)
11. H. Gerard, E. Rivero, D. Six, Basic schemes for TSO – DSO coordination and ancillary services provision (Tech. Rep. No. D1.3). SmartNet project (2016)
12. H. Gerard, E.I. Rivero Puente, D. Six, Coordination between transmission and distribution system operators in the electricity sector: A conceptual framework. Util. Policy **50**, 40–48 (2018). https://doi.org/10/gc49cb

13. F. Gouardères, Internal energy market – fact sheets on the European Union (2018)
14. P. Mallet, P.-O. Granstrom, P. Hallberg, G. Lorenz, P. Mandatova, Power to the people!: European perspectives on the future of electric distribution. IEEE Power Energy Mag. **12**, 51–64 (2014). https://doi.org/10/gdccwr
15. J. Merino, Ancillary service provision by RES and DSM connected at distribution level in the future power system (Tech. Rep. No. D1.1). SmartNet project (2016)
16. L.N. Ochoa, F. Pilo, A. Keane, P. Cuffe, G. Pisano, Embracing an adaptable, flexible posture: Ensuring that future European distribution networks are ready for more active roles. IEEE Power Energy Mag. **14**, 16–28 (2016). https://doi.org/10/gdccxc
17. E.I.R. Puente, H. Gerard, D. Six, A set of roles for the evolving business of electricity distribution. Util. Policy **55**, 178–188 (2018)
18. A. Ramos, E. Rivero, D. Six, Evaluation of current market architectures and regulatory frameworks and the role of DSOs (Tech. Rep. No. D1.2). evolvDSO project (2014)
19. E. Rivero, D. Six, H. Gerard, Assessment of future market architectures and regulatory frameworks for network integration of DRES – the future roles of DSOs (Tech. Rep. No. D1.4). evolvDSO project (2015)
20. E. Rivero, D. Six, P. Mallet, M. Sebastian-Viana, J. Stromsather, M. Baron, evolvDSO: Assessment of the future roles of DSOs, future market architectures and regulatory frameworks for network integration of DRES. Presented at the 23rd international conference on electricity distribution (CIRED), Lyon, France (2015), p. 5
21. S. Ruester, S. Schwenen, C. Batlle, I. Pérez-Arriaga, From distribution networks to smart distribution systems: Rethinking the regulation of European electricity DSOs. Util. Policy **31**, 229–237 (2014). https://doi.org/10/f6wgjq
22. SWECO, ECOFYS, Tractebel Engineering (GDF Suez), PWC, Study on the effective integration of Distributed Energy Resources for providing flexibility to the electricity system (Technical Report) (2015)

Chapter 3
Modeling of Complex Systems Including Transmission, Distribution, Aggregation, Ancillary Services Markets

Mario Džamarija, Guillaume Leclercq, Miguel Marroquin, and Manita Herman

3.1 Introduction

Reducing CO_2 emissions, increasing the share of energy produced by renewable energy sources (RESs) and improving energy efficiency are currently being considered as global targets in order to limit climate change and its consequences. Enforcing these goals has led to a steep growth of RES-based generation, specifically wind farms and solar PV, ever since the 2000s [1]. Solar PV has averaged 42% annual growth over the last decade; onshore wind has averaged 27% [2], and these are the resources that will make up for the largest share of RES utilized in the near future [3]. The expansion in fluctuating, RES-based, electricity generation has led to increased flexibility requirements for the power system. At the same time technological and economic advancements are expected to emphasize the importance of distributed energy resources (DERs), i.e., electric energy storages (EESs), demand response, and distributed generation, in future electric power systems [4]. With an increased deployment of DERs, and a supporting distribution grid ICT infrastructure in place, the flexibility potential of these resources can be utilized to provide services both at local and at system level, to a mutual benefit of DER owners and power system. However, in order to achieve that there is a need to develop coordination schemes between TSOs and DSOs, as discussed in the previous chapter, as well as to implement new market architectures, as detailed in this chapter, so as to manage flexibility offers from DERs.

M. Džamarija (✉)
Danmarks Tekniske Universitet, Kongens Lyngby, Denmark

G. Leclercq
N-SIDE S.A., Louvain-la-Neuve, Belgium

M. Marroquin · M. Herman
Nuestra Nueva Energia Sl, Alicante, Spain

G. Migliavacca (ed.), *TSO-DSO Interactions and Ancillary Services in Electricity Transmission and Distribution Networks*,
https://doi.org/10.1007/978-3-030-29203-4_3

This chapter describes efficient schemes for modeling the different components of the framework described above: aggregation of DER flexibility into market bids, ancillary services (AS) market architecture, market arbitrage, and transmission and distribution network models. The remainder of this chapter is organized as follows: Sect. 3.2 describes the AS market architecture and the transmission/distribution network models, tailored for such market; Sect. 3.3 outlines the aggregation models for DERs; Sect. 3.4 analyzes the connection between the consecutive markets, i.e., market arbitrage, from the aggregator's point of view.

3.2 Modeling of Ancillary Service Markets and Complex Networks

This part describes AS markets and how network constraints are taken into account, Sect. 3.2.1 first discusses the scope of the AS markets that are investigated. Section 3.2.2 covers in detail key market design ingredients: (1) network constraints, (2) timing aspects, (3) market products, and (4) clearing and pricing. Then, Sect. 3.2.3 shows how the AS market clearing problem can be formulated for the different TSO-DSO coordination schemes (CS) described in Chap. 2. Afterward, Sect. 3.2.4 discusses the computational aspects of market clearing algorithms. Section 3.2.5 provides a summary for the AS market setup.

3.2.1 AS Market Scope

3.2.1.1 Market Organization

In order for TSOs or DSOs to procure ancillary/flexibility services, several approaches are possible in practice: a market approach is the most obvious, but mandatory offering or bilateral contracts also exist, for instance [5, 6]. In this chapter, we will focus on the *market* aspect, highlighting what needs to be taken into account when procuring flexibility from DER. In a market, an objective function (e.g., social welfare) is usually optimized, which leads to a fair and efficient behavior of market participants (flexibility providers, flexibility requesters). Traditionally, AS have been provided to TSOs by large centralized power plants and so AS markets have been designed accordingly. In the context of the increasing level of DER, the challenge is to adapt the AS market design to allow DER to compete in a level playing field with the traditional large flexibility providers.

Since DERs are connected to distribution grids, it requires some degree of coordination between the TSO and DSO; otherwise, the grid of the DSO might be jeopardized by the flexibility provision of DER for the TSO, for instance. Chapter 2 focuses on these ways to organize the interactions between TSO and DSO and already tackles some market design aspects, like who (i.e., which *role*) is in charge

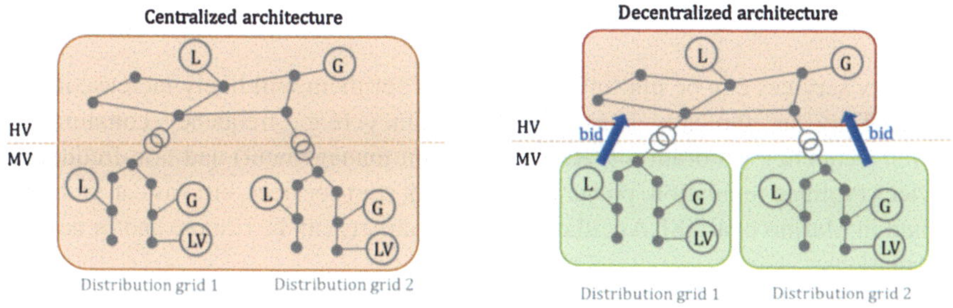

Fig. 3.1 Centralized and decentralized market architectures. (Reproduced from Leclercq et al. [7])

of running the market(s). Also, the market *scope* is intrinsically linked to how the TSO-DSO interaction is organized, through centralized or decentralized market architectures (see Fig. 3.1).

- In centralized market architectures, all flexibility providers offer their flexibility on one single market, irrespective of the grid type (transmission, distribution) or voltage level (high voltage (HV), medium voltage (MV), or low voltage (LV)) they are connected to.
- In decentralized market architectures, flexibility providers bid (i.e., make offers) on the (local) market, depending on their grid location. In these architectures, coordination between markets must be ensured and will be further described in Sect. 3.2.

In terms of market organization, markets can be of different sorts [5, 6], with trades being done over the counter (OTC), i.e., bilateral trades, not centrally organized, or are organized by a central entity (pools). In the framework of AS markets, only markets managed by one or several entities are considered: bilateral trading is typically not very convenient to procure AS at the cheapest cost, except in specific situations (no market liquidity, long-term agreements). Markets centrally organized can also be based on a set of discrete auctions (e.g., daily auction for the EUPHEMIA algorithm coupling the day-ahead market in Europe) or on the basis of a continuous market (e.g., XBID platform for intraday cross-border trading in Europe). Continuous markets are not well suited to AS markets because:

- It is more complex to take network constraints into account (see Sect. 3.2.2.1).
- It would give rise to gambling between nonregulated market parties (flexibility providers) and regulated market parties (i.e., system operators, who need to solve their system/grid issues).

Furthermore, continuous markets are less relevant if the frequency of discrete auctions is high enough (since the main advantage of continuous market is the ability to trade continuously). Therefore, the focus of the next subsections and of further analysis is put on discrete auctions.

3.2.1.2 Ancillary Services Scope

Ancillary services can be multiple and the need for them will likely increase in the future. They are typically clustered in frequency (e.g., frequency containment (FCR), frequency restoration (FRR), congestion management) and non-frequency services (voltage regulation, power quality, black start services, etc.) as can be highlighted in Merino et al. [8]. Ancillary services markets can be reserve and/or activation markets:

- Reserves are purchased by system operator (SO) some time in advance (it can be from several years in advance up to 1 day in advance) to make sure there is enough flexibility capacity available, to be activated if needed in real time. For frequency services, the price of the reserve is typically expressed in (currency/MW/time-period).
- Flexibility is then activated by nature when it is needed, in real time: it can be flexibility from already contracted reserves, but also flexibility coming from providers who haven't sold reserves to the SO. For frequency services, the price of activation is linked to energy (active power) and is usually expressed in (currency/MWh).

The acquisition of AS typically entails different steps, which can be different for different services. These steps include:

1. *Sizing/dimensioning of the reserves* needs for a specific product. For example, a TSO needs to dimension FRR reserve needs, based on some methodology (e.g., statistics of historical data, dynamic sizing [9]).
2. *Procurement of reserves*, i.e., SO purchases the reserves required to fulfill the service. It is usually done through markets, by selecting the cheapest reserve offers according to a merit order (but other possibilities exist like bilateral contracts or mandatory provision foreseen by regulation). This procurement can even be split into multiple horizons: e.g., procure part of the reserve long in advance (e.g., 1 year, 1 month) and another part (more variable) closer to real time (e.g., 1 week, 1 day).
3. *Activation of the services*. This can be done through markets, typically by selecting the cheapest activation offers (according to the merit order). However, for some services, the activation is technical/automated (e.g., activation of FCR is done by a local controller measuring the local frequency and constantly adapting the active power injection based on this).

Another important question is how these markets (procurement, activation) are coupled.

- AS markets can be coupled geographically. As an example, in Europe, there are initiatives [10] to couple the different national replacement reserve (RR) activation markets and also the FRR activation markets.

- Some AS markets can be coupled with energy markets. For instance, balancing reserves and day-ahead energy markets can be co-optimized in a single market session [11], i.e., flexibility providers do not need to account for opportunity costs in the first market cleared since the market will provide the optimal result.
- Different AS markets could be coupled together. For instance, congestion and manual FRR could be combined, provided there is a minimum alignment of market products.

In the following, we do not tackle all these possibilities, but we focus on some of the elements highlighted above (this does not mean that some of the elements presented below are not applicable for the other elements):

- Real-time AS activation market is considered (sizing and reserve procurement are not tackled here, even if very relevant to study): how to design a real-time AS activation market where flexibilities from DER and from assets connected at the transmission network level are activated in an optimal way [7].
- Geographical coupling (i.e., horizontal coupling) is not considered while also an interesting topic. Co-optimization with energy markets is not considered either.
- Service coupling is considered. In particular, the case of combining balancing (like mFRR) and congestion management (both at transmission and distribution level) is considered.

The next section discusses key market design ingredients necessary to set up such real-time AS market aiming at solving balancing and congestion at the same time, using flexibility providers from resources connected to all voltages levels (Fig. 3.2).

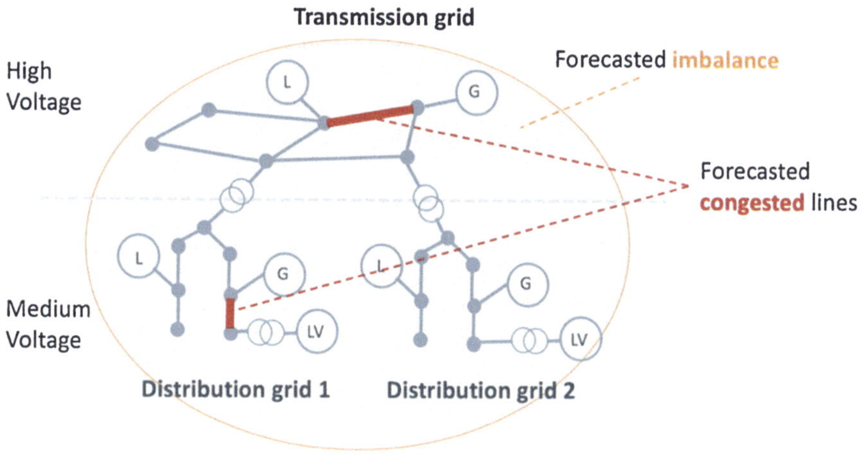

Fig. 3.2 Market aiming at activating services for balancing and congestion (both transmission and distribution) purposes. (Reproduced from Leclercq et al. [7])

3.2.2 Key AS Market Design Ingredients

In this section, we discuss a few key market design ingredients (see Leclercq et al. [7] for more details) that are necessary in order to discuss further the real-time AS activation market in Sect. 3.2.3, for the different TSO-DSO coordination schemes. First, Sect. 3.2.2.1 discusses key AS market timing parameters, which can have a large impact on market efficiency. Then, Sect. 3.2.2.2 focuses on electrical network models to be included in the market clearing algorithm (i.e., the algorithms which decide which flexibility offers are accepted or not and at which price). Afterward, Sect. 3.2.2.3 describes a catalog of (AS) market products that flexibility providers may use to express their offers in the market. Finally, Sect. 3.2.2.4 discusses the objective of the market clearing problem, as well as the pricing aspect.

3.2.2.1 Timing

In discrete market auctions, several key design timing parameters can be defined, as shown in Fig. 3.3.

- The *time horizon* of the market (also called delivery window) represents the time period for which offers are made (e.g., 24 hours for the day-ahead market in Europe).
- The *time granularity* represents a fraction of the time horizon, which allows to have detailed information (market bids, needs, etc.) for each time step of the market time horizon (e.g., 1 hour for the day-ahead market in most European countries).
- The *gate closure time* (e.g., 11:00 AM on day 1 for day-ahead market in Europe) represents the latest time limit for market participants to submit their offers/needs. Gate opening time is also a parameter, less crucial.
- *Market clearing frequency* determines how often the market is cleared (e.g., every day for the European day-ahead market).

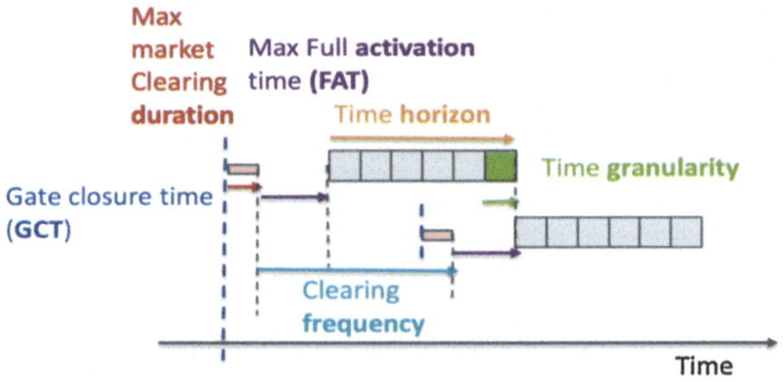

Fig. 3.3 Key market design timing parameters. (Reproduced from Leclercq et al. [7])

- For some AS products, a *max full activation time (FAT)* can be specified (i.e., the time required to answer to a dispatch order + the time required to ramp up/down to the required power level).
- Typically, a maximum time (*maximum market clearing duration*) is allowed to the algorithm to clear the market.

For an AS activation market, the closer the gate closure time (GCT) is to real time, the better, since the forecast accuracy of both TSO and DSO needs and the available flexibility for flexibility providers is improved. However, it is limited by the market clearing algorithm and FAT time. It is also nice to have a long market horizon such that it matches the timescale of the flexibility providers, such that they can internalize their flexibility constraints in the market. However, this is limited by the forecast accuracy that will typically decrease as we go more into the future (e.g., forecast accuracy is typically better on the next 30 minutes than the 30-minute time step in 6 hours time). An interesting way to circumvent this problem is to combine (relatively) long time horizons with much more frequent market clearing, allowing to refine the forecast before every market session. This is done under a rolling horizon framework (see Fig. 3.4), in which market decisions are binding for the first time step and (in most cases) advisory for the remaining time steps of the market horizon.

For a real-time AS market, the choice of these timing parameters is quite critical, since liquidity and forecast accuracy of SO needs may be affected by them, as highlighted in the trade-offs above.

3.2.2.2 Network Models

Another important market design ingredient is the network aspect. Some energy markets typically do not include network constraints at all (e.g., inside a bidding zone in the day-ahead European market) in the market clearing algorithm. It is common practice to then assume that the network model is a *copperplate* (i.e., there are no physical limits). Instead, system operators perform post-processing load flow analyses to check whether results from market are matching the physics. If it is not

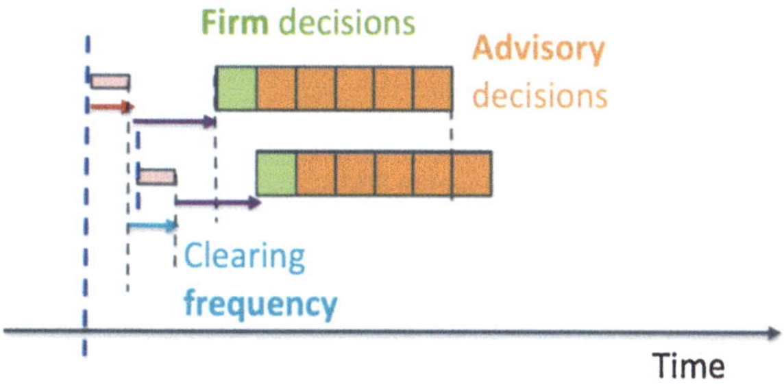

Fig. 3.4 The rolling optimization framework. (Reproduced from Leclercq et al. [7])

the case, they need to perform redispatching/counter-trading (i.e., modify the dispatch instructions) to make sure the physical flows will be feasible in real time. This process may be costly and is not market based and can be quite inefficient [12].

Alternatively, the network constraints could be fully taken into account by considering the full AC (alternating current) power flow equations (which represent the grid stationary constraints quite realistically). In this case, there would be no need to perform redispatching at all since the market results would be fully compliant with the grid physics. However, the big issue is that AC power flows are nonlinear (and non-convex) equations, which make them intractable (in a realistic problem dimension) to manage in a reasonable amount of time in a market clearing optimization problem.

In the framework of real-time AS activation markets focusing on balancing and congestions, it makes sense to consider a detailed grid model for those network areas which are subject to congestions or congestions risks increased by activation of flexibility for balancing purposes. If the grid is strong, then no network model is needed (since no redispatching will be needed afterward anyway). However, since RES and DER are expected to increase a lot in the coming years, it is likely that the networks will be operated closer to their operational limits (instead of reinforcing and oversizing the grid, like the fit-and-forget approach in the past). Therefore, we detail several types of grid models below (with specificities for HV transmission grids and MV distribution grids). The grid model must be accurate enough, to avoid huge redispatching after the market session (quite inefficient), but simple enough to be computationally tractable in an optimization algorithm. Also, the model needs to be aligned with the type of AS which is considered.

Importantly, the network models to be used for transmission and distribution grids are very likely to be different. Even if the AC power flow equations are the same, regardless of the voltage level, there are different aspects which make a network model suitable for transmission, not suitable for distribution, and the other way around, as shown in Table 3.1.

In Geth et al. [13] and Leclercq et al. ([7], chapter 4), the authors describe in details the different types of network models, going from copperplate to full AC power flows. There are two ways to find simplified models of AC power flows:

1. (Convex) relaxation: this process consists in starting from the full nonlinear non-convex model and removing the non-convex parts of the equations. Relaxation

Table 3.1 Some differences between transmission and distribution

Characteristics	Transmission grid	Distribution grid
Topology (structure)	Meshed	Meshed
Topology (operations)	Meshed	Radial
Lines	Inductance > resistance	Inductance ~ resistance
Voltage magnitude margins	Up to 5–6%	Up to 10–15%

models have nice properties, since (1) if a solution is obtained with full model, it will also be a solution of the relaxed model, and (2) if the relaxed problem is not feasible, then the full model is also not feasible.

2. Approximations (which are not relaxations) are another way to approximate the full model. However, it is usually only accurate enough under some specific conditions.

Table 3.2 describes key features for several network models (mostly considered for radial distribution networks), going in decreasing order of complexity from the full AC power flow model to the simple DC model, traditionally used in transmission grid models (PTDF constraints in day-ahead European market actually are built based on the DC model). The intermediate models are the convex SOCP (second-order cone program) DistFlow model, the Ben-Tal DistFlow (i.e., a linearization of the conic constraints of the SOCP DistFlow model) and the simplified DistFlow.

As can be seen, most of them model both active and reactive power (except the DC model), which is important if the market is not supposed to violate operation voltage magnitude constraints. Moreover, line losses are modeled for some of them, but not for the simpler ones. The SOCP DistFlow is quite good and provides physically realistic solution, but the drawback is that a penalty term needs to be tuned to avoid solving problems with "fake losses" (which can be generated because of the relaxed conic constraint). This equally applies to Ben-Tal model. The last lines of the table indicate how the different models perform for solving different grid issues (see more details in Geth et al. [13]; Leclercq et al. [7]).

Table 3.2 Properties of different network models

	AC power flow	DistFlow SOCP	DistFlow Ben-Tal	Simplified DistFlow	DC model
Complexity	Non-convex	SOCP (convex)	Linear	Linear	Linear
Losses modeled	Y	Y	Y	N	N
Reactive power/voltage magnitude modeled	Y	Y	Y	Y	N
Quality	High	High	High	Medium	Low
Tractability	Low	High	High	Very high	Very high
Optimality	Local	Global	Global	Global	Global
Tuning a penalty term?	N	Y	Y	N	N
Undervoltages	Medium	Easy	Easy	Hard	/
Overvoltages	Medium	Hard	Hard	Easy	/
Overcurrent	Medium	Hard	Hard	Easy	Hard

Adapted from Leclercq et al. [7]

3.2.2.3 Market Products

In AS markets, the traded quantities are flexibility offers and flexibility needs. In energy markets, the traded quantities are energy supply offers and energy demand needs (e.g., see day-ahead European market). In AS *reserve* markets, offers are flexibility availabilities (e.g., a capacity of 10 MW to increase injection in the system (i.e., upward flexibility)).

In the scope of this chapter, focusing on AS *activation* markets, offers are typically flexibility energy offers: flexibility providers may offer upward or downward flexibility (compared to a baseline level, as illustrated in Fig. 3.5, agreed beforehand either according to regulation or bilaterally between flexibility provider and system operators [14]).

- An *upward flexibility offer* (see Fig. 3.5, left) represents an increase injection power into the power system, for a certain time period, compared to the baseline case. This increase may be due to increased production and/or reduced consumption.
- A *downward flexibility offer* (see Fig. 3.5, right) represents a decreased injection into the grid, for a certain time period, compared to the baseline case. This decrease may be due to decreased production and/or increased consumption.

Establishing baseline is a challenge (see Brown et al. [14] for more details) for demand response and/or DERs in general (for large power plants and large industrial units, the nominations of the latest energy markets provide an easy baseline everyone agrees on). In practice, DERs providing flexibility also experience rebound/payback effects (see red areas in Fig. 3.5): either this must be coped with by the market participant or it can be internalized in complex market products (see below) if the rebound occurs inside the market horizon.

Figure 3.6 illustrates simple flexibility bids[1], which are typically, in their simplest form, a flexibility quantity (as described above) and a minimum price[2] to provide upward flexibility or a maximum price to provide downward flexibility. A simple bid can be either curtailable (any quantity between 0 and the maximum proposed flexibility quantity can be an acceptable result of the market) or non-curtailable (a flexibility bid can be either not chosen or accepted at full quantity). The latter is important, since it means that the market clearing optimization problem needs to tackle *binary variables*, which typically makes the optimization much more complex. Figure 3.6 also shows that bids can be extended to several pairs of quantity prices, may be either step or piecewise linear, and can be extended to bids over multiple time steps of the market horizon (see Fig. 3.3).

On top of those bids, complex products can be proposed: they are aimed at capturing the dynamics of different flexibility providers while expressing the con-

---

[1] For the sake of the example, only upward flexibility is illustrated. Prices are expressed in cur/MWh.

[2] This minimum price can even be negative for upward flexibility.

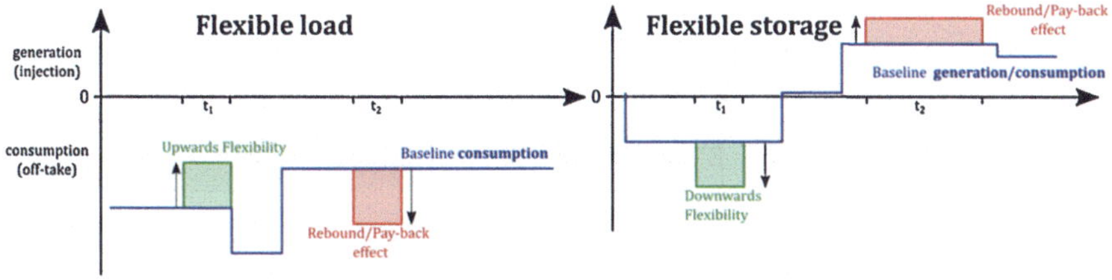

Fig. 3.5 Examples of baseline, activation, and rebound effect profiles of a flexible load and a storage

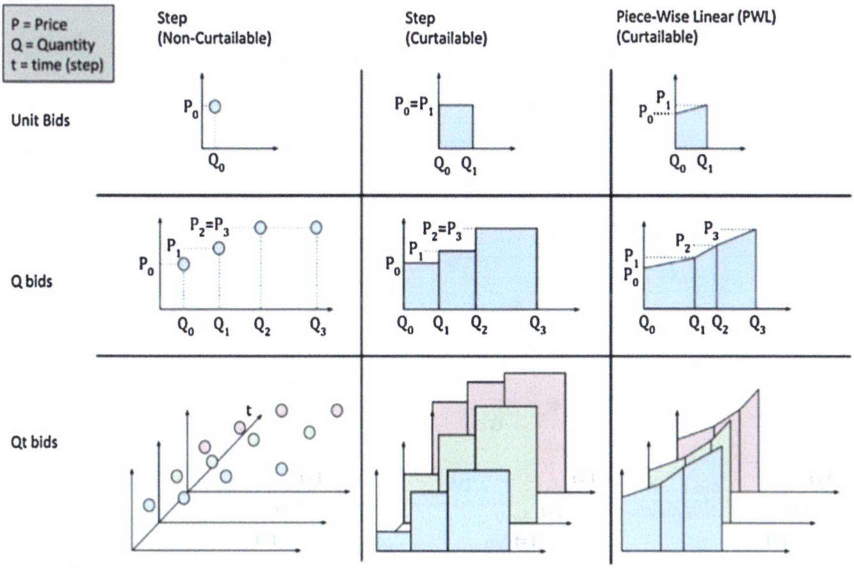

Fig. 3.6 Example of simple bids, over one or multiple time steps, curtailable or not. (Reproduced from Leclercq et al. [7])

straints of assets, aggregators, and system operators. In practice, these bids are expressed as the bids described above, but with additional constraints, as described below:

- *Logical* constraints define logical constraints between the different temporal components of a bid defined over a several time steps of a market horizon or even between two different bids. For example:
  - *Exclusive* constraints prevent two bids from a list of bids to be simultaneously accepted (e.g., may be useful for parallel factory lines).
  - *Implication* constraints to accept a bid B only if a bid A is accepted.
- *Temporal* constraints define constraints between the different temporal components of a bid defined over several time steps. A few examples are as follows:

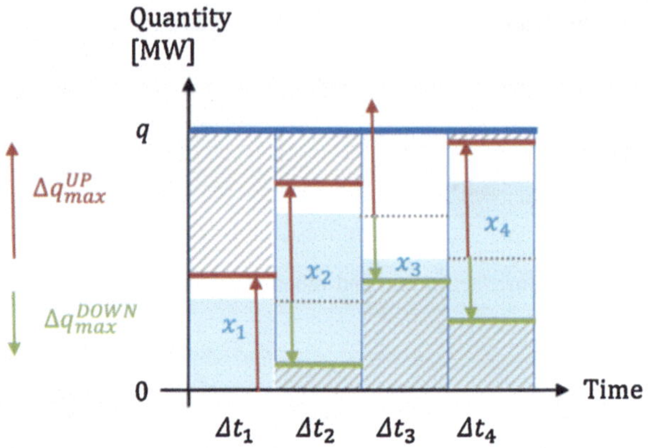

Fig. 3.7 Illustration of ramping constraints. (Reproduced from Leclercq et al. [7])

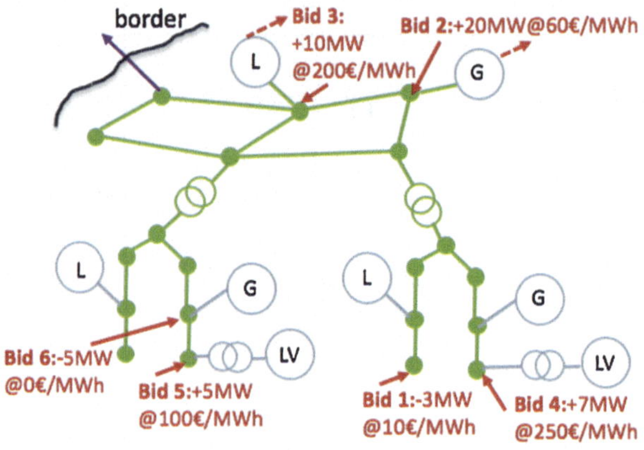

Fig. 3.8 Example of location-dependent upward (+) and downward (−) flexibility bids. (Reproduced from Leclercq et al. [7])

– *Ramping* constraints: limiting the variation of provided flexibility from one time step to the next (see Fig. 3.7). This is a typical constraints for fuel-based power plants.
– *Integral* constraints: set limits on total energy activated on the entire market horizon (may be useful for storage devices).
– *Min/Max* duration of activation.

Importantly, if a network model is included in the market clearing algorithm, the bids must be *location specific* (see Fig. 3.8): *bids must be detailed per node/zone of the modeled transmission/distribution grid.*

In some market designs, system operators need to establish their AS activation needs and express them as bid requests in the AS real-time markets. The bids can be either inelastic (SOs are price-taker) or elastic (SOs also bid a price associated to their AS flexibility needs) if they have other measures to solve their needs. In some other markets, system operators do not bid explicitly but rather provide the market clearing algorithm with a forecast of the network state (i.e., forecast of net power injection (i.e., the local imbalance) at each node of the modeled network: if copperplate, this would be a forecast of the system imbalance).

3.2.2.4 Clearing and Pricing

In an auction, the market clearing problem consists in deciding which orders are (partially) accepted or not and at which price. In mathematical terms, this can be expressed as an optimization problem under constraints (e.g., network constraints, bid constraints). Typically, a common objective is to maximize the social welfare [5, 6]. Other objectives could be to minimize the SO activation costs (it is typically the case when a SO chooses the bids according to merit order, in a single direction (i.e., upward/downward)).

Taking the example of the social welfare (see Fig. 3.9), we can express the problem as:

Maximize *social welfare* such that:

- System is balanced (generation = load).
- Network constraints are respected.
- Bids' constraints are respected.

In terms of pricing, two main approaches may typically be applied:

- *Pay-as-bid* approach: if accepted, market participants receive (pay) the price that they have mentioned in their offer. It is a simple approach, but it does not incentivize market participants to bid at the real cost of flexibility they have.
- *Pay-as-clear* approach: if accepted, all market participants are remunerated at the market clearing price ($p*$ in Fig. 3.9), which represents the marginal flexibility cost (i.e., the cost of the most expensive accepted bid in the case of upward activation).

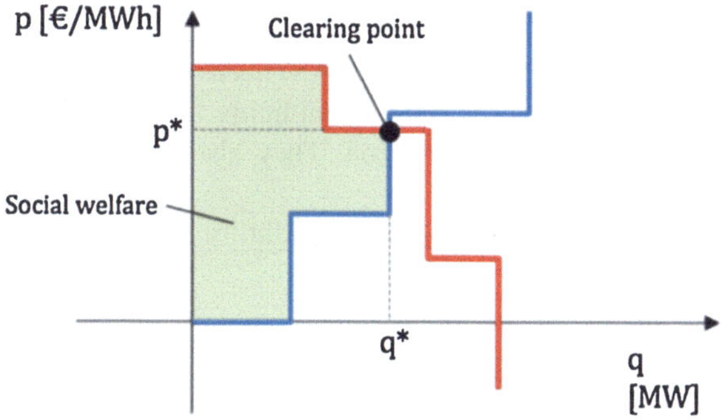

Fig. 3.9 Illustration of social welfare, with optimal cleared quantity $q*$ and clearing price $p*$. (Adapted from Leclercq et al. [7])

Pay-as-clear approach is typically more efficient [15], except in some situations of low liquidity, and is the approach considered in the following text. Using pay-as-clear, and since the network constraints are internalized (to some extent) in the market clearing algorithm, there are typically two approaches for this pay-as-clear approach:

- *Nodal pricing*: a price for the AS is associated to each network node modeled in the problem. This approach allows to price flexibility at its real value. However, it may be quite complex to have a good intuition of nodal prices [16], especially in meshed networks. The nodal approach is typically employed in the USA.
- *Zonal pricing*: a price for the AS is associated to a zone covering multiple network nodes. Prices between zones can be different but must be identical in a given zone. This approach does not allow to reflect the flexibility value inside the zones, but may be useful if the networks inside the zones are strong enough and do not suffer constraints. This approach is typically used in Europe (e.g., day-ahead market coupling).

3.2.3 Market Organization for Different TSO-DSO Coordination Schemes

In Chap. 2, five[3] TSO-DSO coordination schemes (CS) have been proposed (see also Gerard et al. [17, 18]). In this section, we briefly describe how an AS activation market problem may be expressed for each of them (more details can be found in Leclercq et al. [7]). Basically, the different TSO-DSO CS may be categorized in centralized and decentralized architectures, as illustrated in Fig. 3.1 and described in Table 3.3. Note that the common TSO-DSO AS market scheme has two variants: one centralized and one decentralized variant.

Figure 3.10 represents the sequence of some important actions of the different actors (MO (market operator), SO; CMP, commercial market party; FSP, flexibility service provider) for a *centralized AS activation market.*

Figure 3.10 illustrates it for the centralized common TSO-DSO market. For this scheme, the different steps may be expressed as:

1. TSO and DSO send their flexibility (i.e., AS) needs to the MO. It can be explicitly through elastic or inelastic bids or implicitly by specifying the forecasted network state over the market horizon. They also need to send the network constraints.
2. Flexibility providers (CMPs) send their flexibility offers (i.e., bids) before the GCT.

[3] This does not pretend to be an exhaustive list, but a selection of useful ways to cooperate.

Table 3.3 List of centralized and decentralized TSO-DSO architectures

Centralized architecture		Decentralized market architecture	
CS A	Centralized AS market	CS B	Local AS market
CS D1	Common TSO-DSO AS market (centralized)	CS D2	Common TSO-DSO AS market (decentralized)
CS E	Integrated flexibility market	CS C	Shared balancing responsibility model

Adapted from Leclercq et al. [7]

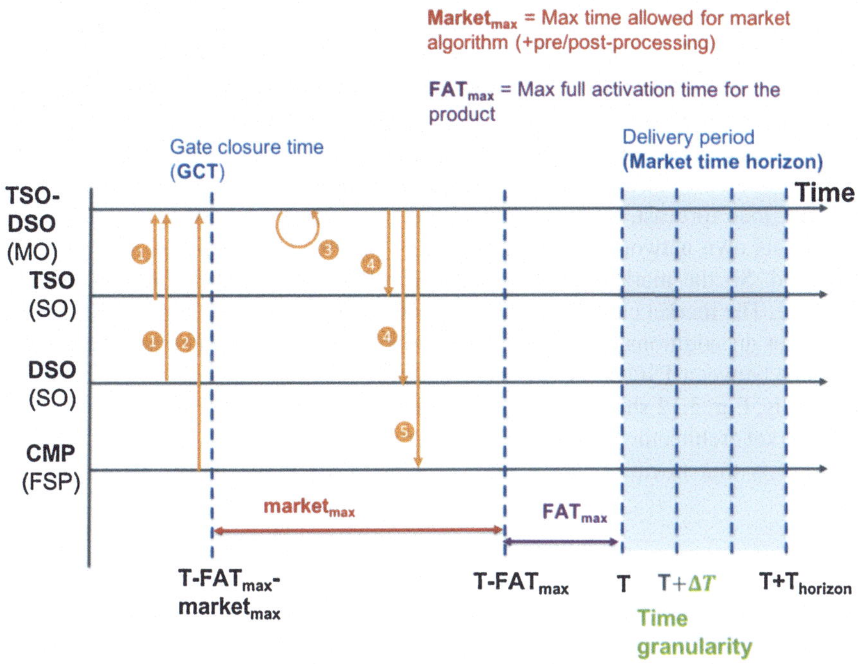

Fig. 3.10 Temporal sequence diagram of actions for a centralized (CS D1) scheme. (Reproduced from Leclercq et al. [7])

3. The MO runs the market clearing algorithm. For instance:
   Maximize *social welfare* subject to:

   – Bid constraints
   – Active (and potentially reactive) *power balance* at *transmission* and *distribution* levels
   – Network *operational* constraints (e.g., line capacities, voltage limits) at transmission and distribution level

4. Results are communicated to TSO and DSOs.
5. Results are communicated to CMPs. Flexibility is dispatched.

The other centralized coordination schemes are quite similar in the steps. For the *centralized AS market*, the DSO is not involved (except in prequalification) and the DSO constraints are not taken into account in the market clearing algorithm. This is a drawback if the distribution grid is operated close to physical limits since costly redispatching actions might be necessary after market is cleared. However, this is good for CMP since they can aggregate flexibility across whole distribution grids. In the *integrated flexibility market*, the steps are quite similar, but an additional actor (CMP as a flexibility requester, e.g., a balance responsible party) submits flexibility needs to the MO, on top of the SO needs.

Figure 3.11 shows the sequence of actions for the *shared balancing responsibility* model (first type of decentralized architecture). In this scheme, TSO and DSOs first agree on a profile of exchange (on active and/or reactive power or voltage, etc.) at each of their interconnection point (i.e., primary substations). This schedule must be established before the GCT: this may be based on latest results from energy markets and/or based on historical data and system/local features (renewable production forecast, load forecast, outdoor temperature, etc.). Then, each SO is responsible to balance its own network, based on flexibility provided by resources located in its own grid. So the markets are cleared in parallel, independently, as indicated in Fig. 3.11. The market clearing problem may be written as above (but applied to each grid), but an additional constraint needs to be inserted to make sure the agreed schedule between TSO and DSO is respected.

Finally, Fig. 3.12 shows the sequence of actions for the other type of decentralized market architecture (see Fig. 3.1), here the Local AS market. The spirit of these schemes is that flexibility from DER is sent to the DSO, which then uses this

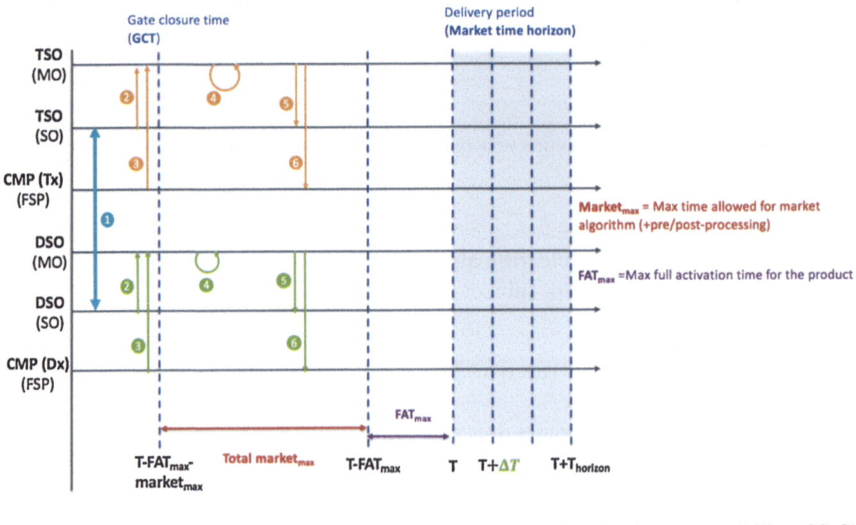

Fig. 3.11 Temporal sequence diagram of actions for the shared balancing responsibility (CS C) scheme. (Reproduced from Leclercq et al. [7])

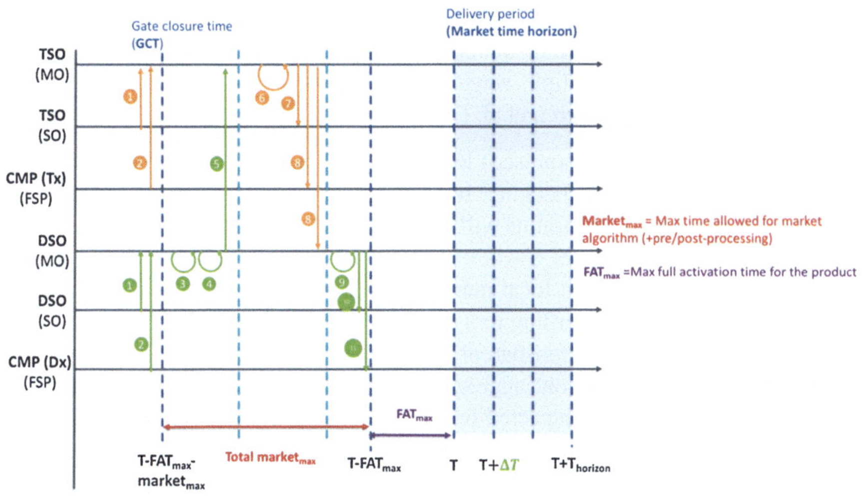

Fig. 3.12 Temporal sequence diagram of actions for the local AS market (CS B) scheme. (Reproduced from Leclercq et al. [7])

flexibility for its own purposes and is responsible to transmit the aggregated flexibility to the TSO for further usage.

The important steps are as follows:

1. TSO and DSO determine their flexibility needs and forecasted network state and send it to their respective market operator.
2. Flexibility providers send their offers to the market where the flexibility is physically linked.
3. The local market operator (DSO) clears the market for its own purposes (e.g., congestion, voltage control).
4. The DSO aggregates the nonused flexibility bids to be sent to the TSO market. How the DSO aggregates and transmits those flexibility bids to the TSO market is further discussed below.
5. The DSO sends the aggregated bids to the TSO market.
6. TSO market is cleared.
7. Market results are sent to the TSO.
8. Market operator dispatches the activated bids of FSP connected to transmission grid and the DSO bids.
9. DSO performs disaggregation of activated DSO bids.
10. DSO (MO) transmits results to DSO (SO).
11. DSO dispatches the activated bids of FSP connected to distribution grid.

In the *decentralized common TSO-DSO market (CS D2)*, step 3 is removed: any local AS need is implicitly solved by only transmitting an aggregated bid offer to the TSO market, which solves the DSO problems, for any accepted quantity.

In CS B and CS D2, the DSO needs to transfer/aggregate bids submitted (see Fig. 3.1, right side) on its own local market to the TSO market. This "smart" aggregation may be done in multiple ways with different levels of complexity (more details can be found in Leclercq et al. [7]):

- *Easy* solution: transmit (nonused) local market bids to the TSO market without any network constraints checking. It is an easy process but there is no guarantee that distribution grid constraints will be satisfied, for the bid quantity accepted by the TSO.
- *Medium* solution: transmit local market bids to the TSO market only if they are prequalified by DSO according to a simple method. But it is not clear how to be at the same time simple, transparent, and fair.
- *Complex* but *smart* solution: aggregate local market bids such that distribution network constraints are respected for any quantity proposed to the TSO market. This involves the resolution of multiple optimization (OPF) problems (see also Papavasiliou and Mezghani [19]).

3.2.4 Computational Aspects

Solving an optimization problem may be difficult and may take some time. For some markets (e.g., long-term reserve markets), the time requirements (i.e., maximum market clearing time) might be loose (i.e., it is not a problem if the algorithm takes tens of minutes or a couple of hours), but when we get closer to real time, the constraints become much more tight. Specifically, for an AS activation market, the less time it takes the better, since it then allows to have a GCT closer to real time (e.g., see Fig. 3.10). However, solving a market clearing optimization problem with many constraints is a difficult problem, whose complexity depends on multiple aspects, dependent on the market design choices (and problem instances), that we can split in factors independent of the TSO-DSO coordination scheme and the factors which are linked to the TSO-DSO coordination scheme.

Among the market design choices impacting the complexity of the algorithm, important factors are as follows:

- The *type of network model*: e.g., copperplate is trivial, while AC model is very tough.
- The *type of market products*: market products with constraints will take more time. But most importantly, market products requiring the introduction of binary (0 or 1 quantities) variables (e.g., simple non-curtailable/non-divisible bids need this feature) significantly complexify the algorithms; indeed, tackling continuous variables or a set of mixed continuous and binary variables impacts the choice of the optimization methods significantly, with the latter being much more challenging.

Table 3.4 Qualitative assessment of the computational complexity of each TSO-DSO coordination scheme

Centralized AS market	Common TSO-DSO AS market (centralized)	Integrated flexibility market	Local AS market	Common TSO-DSO AS market (decentralized)	Shared balancing responsibility model
The **easiest** since only transmission grid	The **most difficult** since full transmission AND distribution grids in a single problem		Optimizations in parallel BUT with smart aggregation using some complexity		Many optimizations in parallel

Reproduced from Leclercq et al. [7]

- the *scaling* effect: complex models may be realistic on small problem instances, but are not realistic when they are scaled up (e.g., size of transmission grid, number and size of distribution networks, number of bids involving binary variables).

On the other side, part of the complexity depends on the TSO-DSO coordination schemes, as shown in Table 3.4. The *centralized AS market* is the less complex, since distribution grid constraints are not taken into account, compared to the other schemes. The most complex schemes are the *common TSO-DSO AS market (centralized)* and the *integrated flexibility market* since there is one single market clearing problem, taking all transmission and distribution grids into account. Regarding the *shared balancing responsibility* model, it is quite simple, since many optimizations are done in parallel, but it requires multiple power platforms. Finally, the decentralized architectures, *local AS market*, and *common TSO-DSO AS market (decentralized)* have intermediate complexities: on the one hand, computations are parallelized by nature, but they require more computations than the shared balancing responsibility model, since they need computation power/time to smartly aggregate local market bids into bids to be sent to the TSO market.

3.2.5 Summary

Section 3.2 described the scope of AS markets and how markets may typically be organized. Then, focus was set on real-time AS activation markets, for which key market design ingredients were introduced: timing parameters, network models and their impact, different types of market products, and clearing/pricing rules. Then we described how the AS activation market model could be formulated for different TSO-DSO coordination schemes, and we, finally, discussed the factors impacting the complexity of the market clearing algorithms in such a context. This chapter does not provide results, but more a methodology on the points to analyze/pay attention to, when designing and implementing market clearing algorithms for AS activation markets, in the context of a coordination between TSO and DSOs.

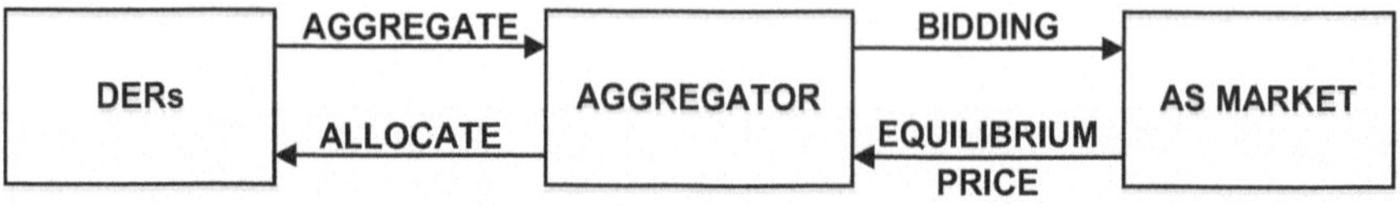

Fig. 3.13 Illustration of aggregation, bidding, market clearing, and disaggregation processes

3.3 Modeling of Small Flexibility Subjects and Aggregation Process Thereof

The majority of DERs cannot compete individually in the electricity markets since a) the power offered to the market must be above a certain threshold, which depends on the geographic location[4], and b) a too high number of market participants may have a negative impact on the performance of the market clearing algorithm. Thus, a new market entity, an aggregator, is needed in order to gather the small sources of flexibility into a single market entity (bid/offer) and get access to the AS market.

Hence, an aggregator plays a key role by making it possible for the small DERs to participate in the AS markets and obtain additional revenue streams while reducing the amount of data passed onto the AS market. Additionally, the risk of imbalance costs, due to DER deviation from the programs established by the market, is taken over by the aggregators. Aggregators are also in charge of the disaggregation process, leading to the resources activation, after the market clearing has taken place. Figure 3.13 shows the aggregator's input and output, i.e., the information flow between the aggregator, individual DERs, and the market clearing algorithm.

In order for a DER aggregator to determine bidding prices – to be submitted to the AS market – it has to know the flexibility provision costs for the DERs in its portfolio[5]. Such costs represent the increases in the DER's cost compared to the base case, when no flexibility is provided (baseline[6] power profile of the DER). We establish the sign convention that a positive flexibility cost represents the minimum amount of money a DER requires to get for providing the flexibility, while a negative flexibility cost represents the maximum amount of money a DER is willing to pay for providing the flexibility.

[4] For example, in the USA, the minimum required size of generation as well as demand response to access the market is 100 kW [20].

[5] Aggregators' bidding strategies will not only depend on costs but also on the gain opportunities offered by the present market session, depending on a guess on the bids from other subjects and on an estimation of the market power owned by the bidder. However, for simplicity we are not going to consider such aspects, which would require a much more complex modeling approach (e.g., game theory-based models).

[6] The baseline can be obtained from the previous market, i.e., day-ahead, intraday, or the previous, i.e., $t-1$, AS market clearing.

3.3.1 Overview of Aggregation Approaches

In the literature, there are different aggregation approaches used for bidding in the electricity market:

- Physical, i.e., bottom-up, approach
- Traces approach
- Hybrid approach
- Data-driven approach

Each of these aggregation approaches has certain shortcomings, either due to the amount of the required input data or due to the accuracy of the represented group of DERs. The aggregation approaches are as follows.

In the *physical, i.e., bottom-up, approach* [21, 22] it is assumed that the aggregator knows all the parameters of each individual device and its real-time status (availability, power set points, etc.). The physical approach studies the bids from the perspective of individual physical entities, hence the name bottom-up. The physical approach can become hard to implement when many heterogeneous energy resources are included, since different input parameters and constraints have to be obtained in order to represent the group of devices accurately. In that case an alternative, either hybrid or data-driven, approach can be considered. The advantage of the physical approach is straightforward disaggregation.

The *traces approach* [23, 24] shares similarities with the physical approach since it takes account of the individual devices in the aggregation process. The exception is that the flexibility providers are characterized by load profiles, and the cost associated with each of the profiles, rather than by the exact physical DERs' characteristics. This could be convenient, for example, when due to confidentiality reasons, prohibitive complexity, or insufficient accuracy of the available models, physical characteristics can't be provided. The aggregation is represented by all the possible combinations of the feasible device profiles. As it is the case for the bottom-up approach, the particular advantage of the traces approach is that the disaggregation becomes trivial. When a bid is formed from a particular combination of the feasible device profiles, it already contains the information which device needs to change its schedule, if the bid is accepted.

The *hybrid approach* [25, 26] uses a single or a limited number of virtual devices in order to represent the overall group of aggregated devices. Such practice reduces the number of individual devices and avoids the large number of input parameters required for the aggregation models. Hence, it can be argued that, in the case when a large number of devices needs to be aggregated, the hybrid approach seems to have an advantage over the bottom-up approach. The drawback of the hybrid approach is that, in case of heterogeneous devices, it introduces a modeling error, since it represents the entire group of aggregated devices by the parameters of a single or a limited number of virtual devices. A way to reduce this error is to cluster the devices that have similar model parameters, such that there are homogeneous devices in each cluster. As the number of clusters increases, the hybrid approach becomes closer to the bottom-up approach. In the case when the number of clusters equals the number of individual devices, the hybrid approach becomes the physical, bottom-up, approach.

In the *data-driven approach* [27–29] the physical entities, and their specific technologies, are not considered any longer, as the behavior of DERs is analyzed at the group level. The availability of the relevant recorded data is fundamental for this approach, since it intends to emulate and predict the behavior of a group of devices. In contrast to the physical approach, the data-driven approach does not require any input parameters of devices, either from the literature or the practical experience, since it is carried out by using a more accurate level of information in case real historical data is available. Due to this reason, it requires more input data than the physical approach, which can be problematic in case of data scarcity. Unlike the other aggregation approaches mentioned here, the data-driven approach requires a disaggregation model.

Table 3.5 provides an overview of the references and characteristics of the disaggregation for each of the approaches.

Figure 3.14 illustrates aggregator's generic bid function for bidirectional flexibility, showing positive and negative energy volume blocks, each offered at a different price. The abovementioned aggregation approaches generate a bid curve, obtained by the horizontal summation [30, 31] of the individual bid functions. The horizontal summation implies the addition of the bid blocks' energy volumes while sorting them, according to their bid price, in the ascending order, effectively arranging them from the lowest to the highest price, which is commonly known as a bid stack [32]. Figure 3.15 illustrates the horizontal summation of bids, for an arbitrary time step; the left-hand side of Fig. 3.15 shows two individual bids for bidirectional flexibility; the right-hand side of Fig. 3.15 shows a "U-shape" bid curve. The "U-shape" bid curve entails that the aggregator is offering to deviate from the committed day-ahead/intraday/AS market_{t-1} market baseline, at rising cost. The aggregator will accept all of the market clearing prices equal to/above the bold bidding curve, shown on the right-hand side of Fig. 3.15.

As previously described in Sect. 3.2, we assume nodal pricing is applied in the AS market. Due to this, the aggregation is done separately at each MV distribution grid node. The MV nodes, at which the aggregation is conducted, are illustrated in Fig. 3.16, in which they are indicated by the dashed lines.

Table 3.5 Overview of different aggregation approaches for DERs

Aggregation approach	References	Disaggregation
Physical	[21, 22]	Straightforward
Traces	[23, 24]	Straightforward
Hybrid	[25, 26]	Straightforward
Data-driven	[27–29]	Model

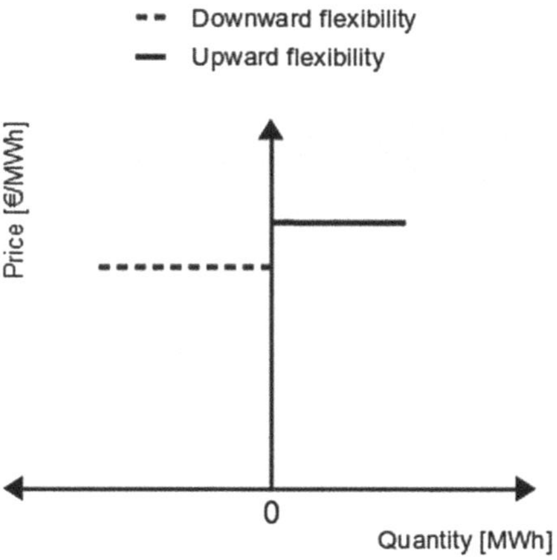

Fig. 3.14 Upward (solid line) and downward flexibility (dashed line) generic bid function

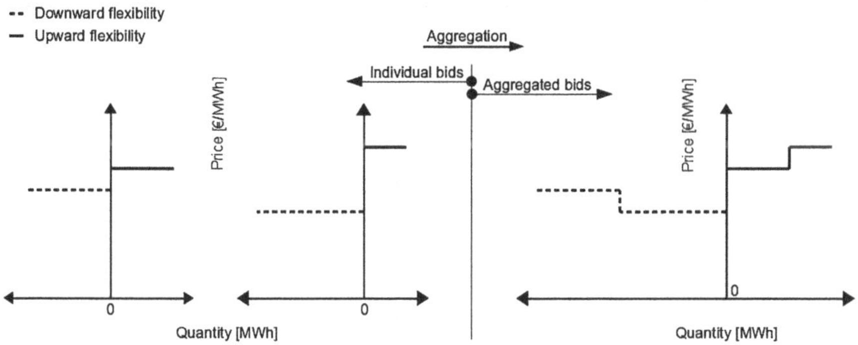

Fig. 3.15 Aggregation of individual bids by horizontal summation, into a cumulative bid curve, including upward (solid line) and downward flexibility (dashed line), in reference to the baseline

3.3.2 Aggregation Models

The aggregation of DERs has been previously considered in [33–35]. In [33] a single aggregation model is used for demand response and distributed generation, located at the medium voltage distribution level. This aggregation model uses a bottom-up approach and it represents aggregated DERs, plus the local network constraints, as a single entity at the transmission level-distribution level interface. This is the concept of a virtual power plant (VPP) which aggregates the capabilities of heterogeneous DERs, at the same geographical location, into a single capability diagram, in the same way a large-scale generator is presented to a TSO, with its PQ capability[7] and a production cost. The methodology describing how aggregated PQ capability and redispatching costs are calculated is detailed in [33].

[7] PQ capability diagram characterizes active and reactive power limits of a generator.

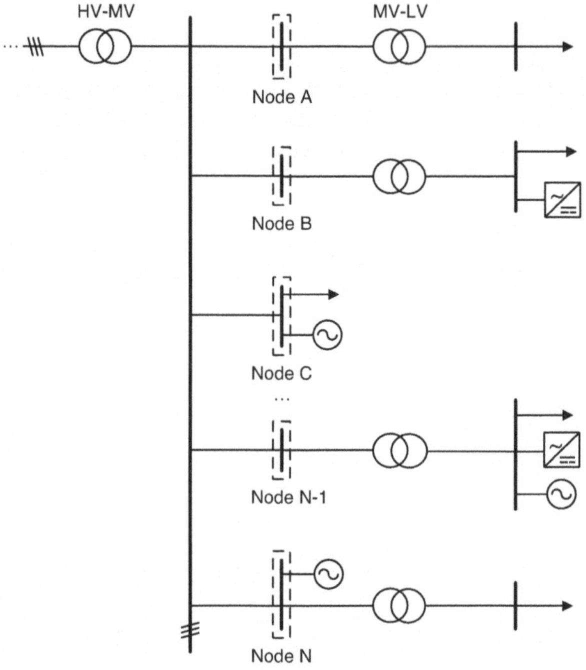

Fig. 3.16 Distribution network illustration

The aggregation of household devices, some of them controllable[8], is considered in [35]. The aggregator clusters the consumers sharing similar characteristics in terms of appliances' usage and type, geographical location, etc. The aggregation was tested with data comprising five clusters. The aggregator sends a price and volume signal to each cluster expecting a certain response and having as objective its revenue maximization. Each cluster then takes this signal and optimizes the response of the devices, with the objective of minimizing the end users' electricity bill. This kind of a complex disaggregation adds latency in response to the market clearing, which is still acceptable for the day-ahead and intraday markets, considered in this work.

However, when considering near-real-time AS markets, complex aggregation and disaggregation process is to be avoided. Complex disaggregation, akin to the one in [35], tends to add latency in response to the market clearing. Hence, the aggregator here, unlike the one considered in [35], uses several technology-specific aggregation models, aimed at separate DER categories, in order to take into account the physical constraints of the devices being aggregated while enabling a fast, straightforward, aggregation/disaggregation procedure. Due to

[8] Here, we consider thermostatically controlled devices, such as air-conditioning and heaters, as well as time-shiftable devices (i.e., controlled by a timer) like washing machines and dishwashers.

Table 3.6 Aggregation approaches used for aggregation of different DERs

Aggregation model	Aggregation approach
CHP units, curtailable generation and curtailable loads, EES units, TCLs	Physical
Atomic loads	Traces

this reason only the DERs that are reasonably similar in terms of their specific core features are grouped together in the same aggregation model. Hence, in the aggregation process, stationary EES and EVs are together considered as EES units, curtailable generation and curtailable loads are grouped into a unified model, wet appliances and industrial processes are grouped into atomic loads, all of the TCLs are considered in TCL model, and CHP is considered in the CHP model. The bottom-up approach is chosen as the preferred option for the aggregation models due to the lower number of devices, which are being aggregated into each MV node, and the physical characteristics of the aggregated devices. However, due to the specific nature of some of the DERs, namely, wet appliances and industrial processes, which cannot be paused/interrupted once started, traces approach is used in one of the aggregation models. This is shown in Table 3.6.

The grouping of DER categories, done according to the individual models' constraint similarities, simplifies the bidding system. As the market clearing mechanism is able to cope with multiple bid types, belonging to different aggregation models and originating from the same aggregator, the simplest approach is for the aggregator to allow all five aggregation-type-specific categories, mentioned above, to generate bids for their own aggregated devices. That is to say, a single aggregator can use several aggregation models for providing its bids. By doing so, every bid that is accepted by the market can then be assigned to the corresponding device-type-specific disaggregation algorithm, which is best equipped for optimally distributing the allocated flexibility over its individual devices. The reason behind this approach is the fact that it would not be convenient to build an overarching aggregation model, as it would inevitably make the disaggregation complex; i.e., it would become unwieldy to trace the accepted bids back to the individual devices. Once the market is cleared, the aggregators are notified whether the submitted bids are accepted or not. Essentially, the accepted bids are the ones with the bidding price equal/lower in reference to the market clearing price. Hence, the identification of DERs, which are going to be activated for the flexibility purpose, is straightforward. As previously mentioned, the aggregator is obliged to provide the accepted flexibility quantity to the market, or otherwise it is subject to imbalance costs. In case a bid is partially accepted,[9] if the market provides such an option, the optimal disaggregation approach is to activate the lowest possible number of units. By doing so, unnecessary activation costs, of more than required devices, are avoided. Despite the

[9] In the literature it is referred to as the curtailable bid, which has a single price for a range between two quantities, and the market operator can accept any value between these two quantities.

aforementioned advantages, the drawback of the aggregation approach described here, compared to the one in [33], is the higher number of bids going to the market, since five relatively simpler aggregation models are used, instead of a single complex one.

3.4 How Day-Ahead and Intraday Markets Influence the Bidding, How the Aggregator Can Benefit from Speculation in the Market with Different Time Horizons, What Are the Risks Involved

In increasingly complex energy markets, supply and demand are satisfied in a series of consecutive markets. These markets respond to different needs and constraints, starting from futures markets that serve to hedge price risk up to several years ahead till real-time markets, dispatching unit flexibility. Between the first and the last markets we find a sequence of markets, as follows:

- Day-ahead (or spot) market: The day-ahead market takes place 1 day before the actual delivery of electricity in the form of supply and demand auctions for every hour or quarter hour. Typically, the day-ahead market closes around noon. It is the most significant market, in terms of volumes traded, compared to the intraday and the ancillary market. Volume matching and price setting are performed jointly between the European energy markets under the flow-based coupling mechanism, subject to complex constraints. Shortly after 12 noon, the prices for each hour and balancing zone are set and the day-ahead positions (nominations) are communicated to producers and consumers. These prices and positions set the day-ahead baselines for generation and consumption.
- Intraday market: Since the day-ahead auction takes place hours before delivery, market operators typically arrange the intraday market to allow for trading of electricity closer to the period of delivery, or the so-called gate closure (e.g., EPEX SPOT markets) [36–40]. On exchanges and broker platforms, market participants exchange bilaterally and continuously volumes of electricity (standard traded products or contracts), such as baseload delivery for every hour, half hour, or quarter hour. Some market operators, e.g., in Spain and Italy, adopt different market mechanisms, for example, a series of intraday auctions. The intraday market serves as a platform to balance the differences between the day-ahead forecasts for production and consumption and the intraday forecasts, thus signaling different supply and demand equilibriums (noon of D-1). The prices and volumes settled in the intraday market provide the baselines for intraday adjustments. Participation in the intraday market is not yet fully compulsory in all European countries.
- Ancillary market: With the advancement of the electricity markets, TSOs have developed the ancillary (balancing) market to enhance transparency, openness, and structure, bringing opportunities for additional revenues to flexible assets.

Ancillary markets are activated between the intraday gate closure and the real-time delivery. They are usually steered by TSOs, which have the most updated information on the actual grid status and which operate grids in a secure and cost-efficient manner. Ancillary markets are organized in the form of auctions where flexible producers and consumers bid their ability to redispatch volumes from their intraday baselines in exchange for a profit. For example, a consumer will offer to curtail his consumption thereby increasing volumes to the system (upregulate) if the price it will be paid to do so is significantly higher than the price it paid for the volumes in the previous markets. On the contrary, a generator would offer to curtail his production, thereby removing volumes from the system (buyback or downregulate), if he is able to buyback the production paying a significantly cheaper price than the price received in the previous markets.

The AS market, described in this chapter, takes place in the ancillary market near real time, therefore comes after, at least, two very important markets that have conditioned the positions of the market participants.

Typically, flexibility has two symmetrical sides: curtailing (upregulation) and consuming (downregulation). A flexible asset can curtail consumption if it is scheduled for a curtailment, or it can continue consuming energy. As a result, a flexible asset would reach a market auction with flexibility in one direction, depending on its choice to curtail or consume in the previous market. In other words, the choice of a DER to make itself available for up- or downregulation in the AS market is constrained by its choice in the previous market. A DER will only be able to deliver up- or downregulation based on its relative value to the previous market, as well as the expectations of future value for activations.

As stated above, the sequence of markets, comprised of the day-ahead, the intraday, and the AS market, represents a sequence of opportunities to valorize DER flexibility. This effect works in two ways, which are significant to our modeling exercise. First of all, the AS market comes last when deciding how to act in real time. This means that the aggregator's decision is constrained by the flexibility direction (up- or downregulation) relative to its previous actions. Secondly, in the case of complex flexibility provided by the demand assets, it often happens that an *activation* in one period has an impact on the available flexibility in future periods. This is the case, for instance, in some battery storage configurations, cogenerations, TCLs, wet appliances, etc. There could be a limited number of activations per day, or there could be integral constraints that render the choice to activate now into the impossibility to activate later, thereby giving up on the future value of the flexibility within a given time horizon. Since the current value of the future energy delivery is available in a transparent intraday market, DERs and aggregators face a choice between activations, hence between current real-time (AS market) prices and current intraday prices or future intraday/AS market prices.

It is worth mentioning that the flexibility is not necessarily symmetric, such as in the case of renewable assets. The renewable generation technology relies on resource availability, which is variable and unpredictable by nature. It is generally accepted that the renewable assets can provide downregulation (curtailment of production), but their ability to provide upregulation is way more controversial, since the resource availability in the future is uncertain. Hence, the aggregator usually considers only downregulation in its bidding strategy, for this asset category.

Marginal costs are considered in the aggregator's decision to bid. The objective of DERs' schedule optimization is the rational maximization of the available resources (e.g., maximized techno-economical value of available activations), constrained by the number, frequency, latency, etc., of activations. In the SmartNet project (http://smartnet-project.eu/), we have proposed to deal with this paradigm by incorporating a dummy cost named market discomfort cost (MDC), into the marginal cost, which accounts for the following effects:

- MDC value will increase when intraday prices for future delivery are almost as interesting or when the market volatility is high. This reduces the probability of activation now. However, this does not mean that the aggregator will reject activations, only that it will, at least, require making profit in order to be satisfied.
- MDC value will decrease when the intraday prices for future delivery are far less interesting or when the market volatility is low. This enhances the probability of activation, yet it will result in a larger indifference from the aggregator to be activated now or later.

In essence, the MDC works as an opportunity cost of the future activations. This consideration has two implications regarding MDC:

- It is always positive or zero.
- It is proportional to the expected value of future activations.

Since different aggregators, just as different people, have different expectations, risk aversion behaviors, and market information, it is assumed that leaving this MDC flat for every aggregator in every node would not be realistic and that it would lead to a binary response from aggregators across the grid; i.e., they would all switch at the same prices, resulting in large capacity steps.

Instead, we opted to use a weighting parameter multiplying the risk premium, described above, by a randomizer parameter omega ($\omega \in [0, 1]$). This way, each aggregator in each node will have its "own" risk aversion level, while all of them will increase or decrease when volatility fundamentals go up or down, respectively. In the simulation the omega parameter is assigned to each aggregator and for each node once and then kept constant throughout the simulation exercise.

Figure 3.17 presents the case of a consumption asset whose marginal cost to curtail is above the day-ahead price for a given period; hence the asset has no curtailment programmed. At the present point in time, the intraday price signal anticipates an event in the network that should push prices up. The situation drives intraday value for those future hours up which suggests bidding in the AS market. In an efficient market, the intraday prices would signal the issue. Then, the market players would react and we would see how the intraday prices take into consideration the updated view on the situation.

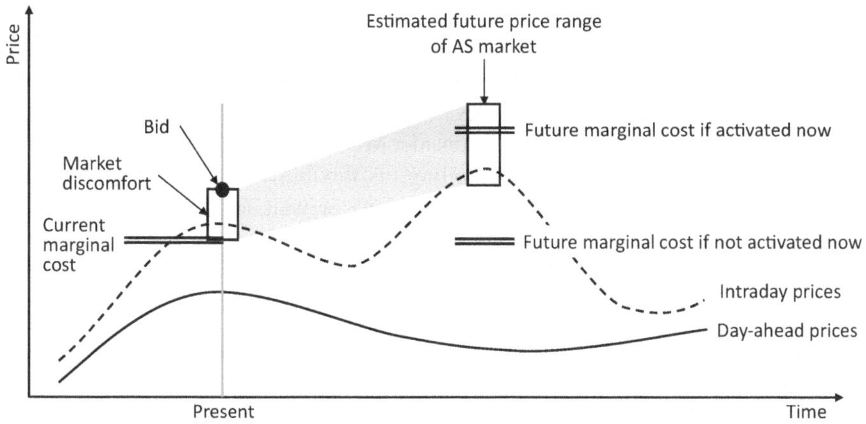

Fig. 3.17 Representation of the market discomfort cost

The aggregator forms a vision of the expected clearing price for upward flexibility (load curtailment) in the future. Provided that the number of curtailment activations is limited, its marginal activation cost might increase for the future activation, provided that it is activated now. Hence, the aggregator will consider a risk premium that takes into consideration the possible future market-related variables, as well as its own forecast error. These will be reflected in MDC, which is aggregator and node dependent:

$$c_t^{\text{marketdiscomfort}} = T_{ij} + \omega \cdot \text{RiskPremium}, \ \forall t, \tag{3.1}$$

where:

- $c_t^{\text{discomfort}}$ quantifies the market discomfort cost at time step t.
- T_{ij} represents the difference between the estimated/forecasted market clearing prices of the consecutive markets.
- RiskPremium is a function of the next market clearing price uncertainty, current market clearing price uncertainty, bid volume uncertainty, expectation of future activation, etc.
- ω is a nonnegative coefficient, $\omega \in [0, 1]$, that weighs risk aversion, whereby $\omega = 0$ represents the extreme case of a risk taker, while $\omega = 1$ represents the extreme case of risk aversion.

MDC is incorporated in the existing flexibility cost[10], in order to produce a bid in the present AS market, according to:

$$\begin{aligned} c_{DER,t}^{\text{flex}} = c_{DER,t}^{\text{discomfort}} + c_{DER,t}^{\text{indirect}} + c_{DER,t}^{\text{operational}} \\ + \Delta_{DER,t}^{\text{revenue}} + c_t^{\text{marketdiscomfort}} \end{aligned} \tag{3.2}$$

[10] For simplicity we omit the detailed equation for each component of the flexibility cost equation. A comprehensive explanation, with equations, can be found in deliverables D2.1 [41] and D2.2 [42], of the SmartNet project.

If a bid that considers MDC is accepted, the additional revenue from MDC should compensate for the opportunity cost received from activation in the nearby future. In other words, MDC represents an artificial cost that makes the aggregator indifferent between an immediate activation and the one in the future at a (potentially) better profit. If the bid does not consider MDC, the aggregator will act accordingly and place its choice either on selling his flexibility on the intraday market (which pays better than the current AS market) or wait for a future AS market auctions.

At this point, the behavioral simulation of DERs and aggregators acquires a new dimension of complexity. We approach this in two stages:

1. Our simulation environment runs a day-ahead and an intraday price simulation, which outputs a net regulating volume and aggregator-specific intraday baseline scenarios that assume a rational and economically driven optimization of their flexibility.
2. Inside the AS market simulation environment, we have postulated a way to cope with this new challenge through a MDC. This new cost parameter is meant to account for the potential inhibition of the aggregators, an element to make the aggregators refrain from jumping too fast at the first opportunity, disregarding the future value of a limited number of activations available. Hence, the MDC makes the aggregators refrain from offering their flexibility at a purely technical cost and introduces a risk premium, or a psychological factor, that increases the required return from activation, thereby generating the right economic incentive to the aggregator and ultimately DERs (Figs. 3.18 and 3.19).

Example of Arbitrage (Solar)

To illustrate, we give an example of an electricity market in 2030. Let's assume there is a significant share of installed renewable capacity and that it is predominantly of PV nature.

The day-ahead auction is based on the expectation that the resource will be highly available. Under this circumstance, we expect depressed prices in the central hours of the day due to the very high forecast of cheap renewable generation. This results in the scheduling of some flexible consumption in the hour of the lowest price, which is the hour with the highest renewable share in the generation stack; let's assume it is the hour ending at 13 hours (H13).

Following the day-ahead auction, the intraday market opens. Here, the price for H13 is sensitive to the updated forecast on cloud coverage at around noon to 13 hours. Since the day-ahead PV generation forecast was high, resulting in low day-ahead prices, the potential opportunity for the DER comes from the increasing prices in the intraday market, owing to the forecast of unforeseen cloud coverage and therefore lower-than-expected PV infeed.

As we know, our DER has placed its flexible consumption in H13 to buy at a cheap day-ahead price. An increase in price represents an opportunity for the DER to resell its cheaply acquired volumes at a profit, thereby scheduling a consumption curtailment in H13. Let's assume this is the DER's choice at the gate closure.

When considering participating in the AS market, the DER's updated baseline is to curtail consumption, and its flexibility is to consume again, i.e., to buy cheap energy. The DER will consider the option to bid in the AS market in downregulation.

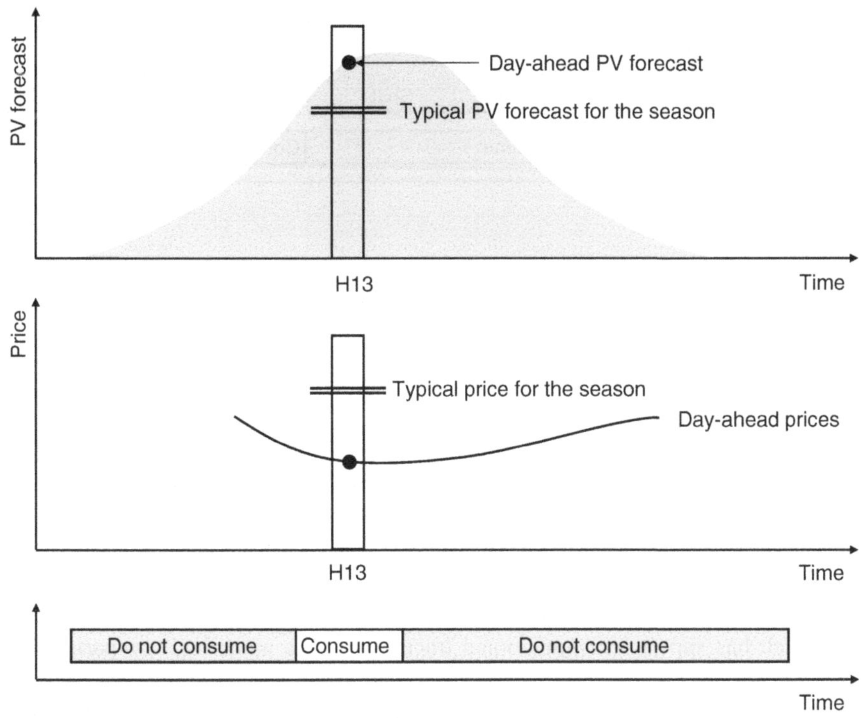

Fig. 3.18 Example of arbitrage for solar PV – base case

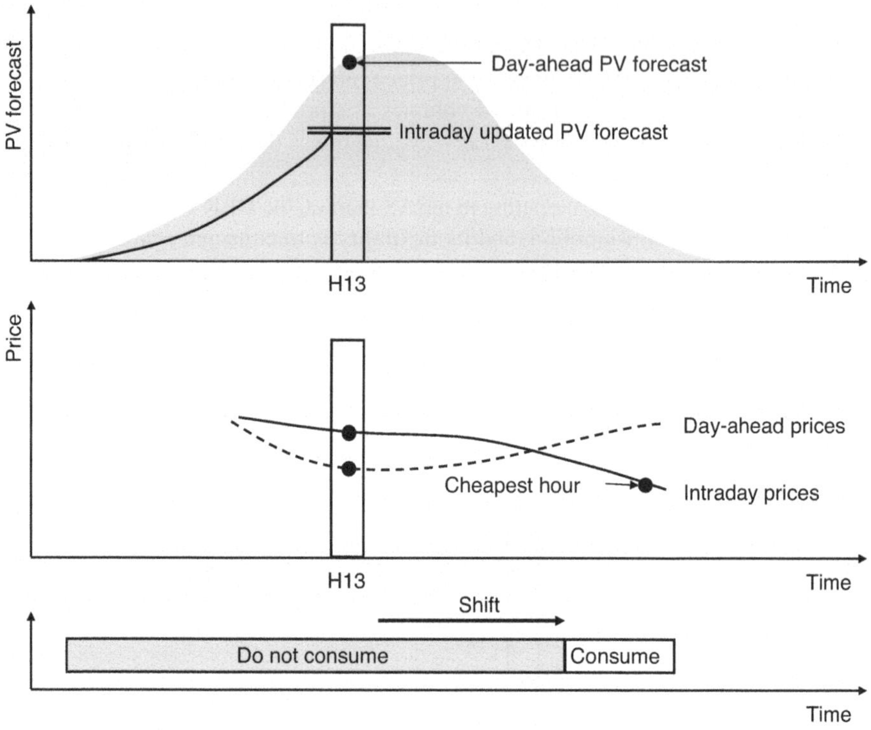

Fig. 3.19 Example of arbitrage for case of solar PV – updated forecast

3.5 Conclusion

Cost-optimal scheduling of DERs, i.e., market clearing, is done while taking account of the distribution network's physics. Since optimal power flow (OPF) is a nonlinear and non-convex problem, and since integer variables exist within the market clearing constraints, this originally represents a MINLP problem, which is difficult to solve. A limited timeframe for clearing the AS market, and the need to consider an accurate network model, calls for a convex approximation of the power flow. This approximation allows to develop an overall model which is mixed-integer convex and which has superior computational tractability. The numerical comparison of OPF formulations shows that the second-order cone programming branch flow model offers high computational tractability, as well as accuracy. Another key message is that the objective of the market is highly affected by several pre-defined conventions, such as types of bids allowed, price-quantity convention, network model complexity, and the type of objective. Furthermore, parameter values can affect both the behavior and the performance of the clearing algorithm. Some important ones include the duration of time step, the length of prediction horizon, and the frequency of clearing.

The key roles of an aggregator are as follows:

1. Enabling DERs to access the market
2. Reducing the amount of data passed to the market clearing algorithm
3. Taking the risk of the imbalance cost upon itself
4. Activation of resources after a successful market clearing

Regarding the aggregation process, the more distinct the features of the aggregated devices are, the less accurate the approximations can become during their aggregation and the more difficulties can arise during the disaggregation stage. Therefore, in the presented AS market it has been postulated that the aggregator will use separate aggregation models, depending on the types of the devices being aggregated and the data availability thereof. The market clearing mechanism is able to cope with multiple bid types, belonging to different aggregation models and originating from the same aggregator. Due to that, the simplest approach for the aggregator is to use a technology-based aggregation, where DERs with similar technical constraints have been grouped into one aggregation model (e.g., stationary EESs and EVs are grouped together). This is a step toward a source specific AS (retail) market.

Arbitrage is introduced into aggregator's bidding strategy in a form of an artificial cost, named MDC. It is intended for preventing from bidding at a purely technical cost, when there is an opportunity to earn more in the near future, from a limited number of available activations.

References

1. REN21, Renewables global status report 2012 (Jun 2012). [Online]. Available: http://www.ren21.net/. Accessed 26 Sept 2018
2. Energy technology perspectives 2012: Executive summary. [Online]. Available: http://www.iea.org/publications/freepublications/. Accessed 26 Sept 2018
3. C.J. Barnhart, M. Dale, A.R. Brandt, S.M. Benson, The energetic implications of curtailing versus storing solar- and wind-generated electricity. Energy Environ. Sci. **6**, 2804–2810 (2013)
4. F. Martín-Martínez, A. Sánchez-Miralles, M. Rivier, Prosumers' optimal DER investments and DR usage for thermal and electrical loads in isolated microgrids. Electr. Power Syst. Res. **140**, 473 (2016)
5. D.R. Biggar, M.R. Hesamzadeh, *The Economics of Electricity Markets* (John Wiley & Sons, Chichester, 2014)
6. D. Kirschen, G. Strbac, *Fundamentals of Power System Economics* (John Wiley & Sons, Chichester, 2004)
7. G. Leclercq, M. Pavesi, T. Gueuning et al., Network and market models. SmartNet project, D2.2 (2019)
8. J. Merino, I. Gómez, E. Turienzo, C. Madina, Ancillary service provision by RES and DSM connected at distribution level in the future power system. SmartNet project, D2.1 (2016)
9. K. De Vos, N. Stevens, O. Devolder, et al., Dynamic dimensioning approach for operating reserves: Proof of concept in Belgium. Energy Policy **124**, 272–285 (2018). https://doi.org/10.1016/j.enpol.2018.09.031
10. ENTSO-E, Balancing and ancillary services markets. https://www.entsoe.eu/about/market/#balancing-and-ancillary-services-markets. Accessed 1 Jun 2019

11. M. Caramanis, E. Ntakou, W.W. Hogan, et al., Co-optimization of power and reserves in dynamic T&D power markets with nondispatchable renewable generation and distributed energy resources. Proc. IEEE **104**, 807–836 (2016). https://doi.org/10.1109/JPROC.2016.2520758
12. P. Holmberg, E. Lazarczyk, Congestion management in electricity: Nodal, zonal and discriminatory pricing. IFN Working Paper No. 915 (2012)
13. F. Geth, R. D'Hulst, D. Van Hertem, Convex power flow models for scalable electricity market modelling. CIRED – Open Access Proc. J. **2017**, 989–993 (2017). https://doi.org/10.1049/oap-cired.2017.0325
14. T. Brown, S. Newell, D. Oates, K. Spees, International review of demand response mechanisms (2015)., 83
15. S.F. Tierney, T. Schatzki, R. Mukerji, *Uniform-Pricing Versus Pay-as-Bid in Wholesale Electricity Markets: Does it Make a Difference?* (ISO, New York, 2008)
16. A. Papavasiliou, Analysis of distribution locational marginal prices. IEEE Trans. Smart Grid **9**, 4872–4882 (2017). https://doi.org/10.1109/TSG.2017.2673860
17. H. Gerard, E. Rivero, D. Six, Basic schemes for TSO-DSO coordination and ancillary services provision. SmartNet project, D1.3 (2016)
18. H. Gerard, E.I. Rivero Puente, D. Six, Coordination between transmission and distribution system operators in the electricity sector: A conceptual framework. Util. Policy **50**, 40–48 (2018). https://doi.org/10.1016/j.jup.2017.09.011
19. A. Papavasiliou, I. Mezghani, Coordination schemes for the integration of transmission and distribution system operations. 2018 Power Systems Computation Conference (PSCC), IEEE (2018), pp 1–7
20. A. Levitt, Comparing RTO market framework rules in DER contex (Aug 2016). [Online]. Available: http://www.pjm.com/~/media/committees-groups/committees/mrc/20160824-special/20160824-item-02-der-rto-benchmarking.ashx. Accessed 26 Sept 2018
21. M. Parvania, M. Fotuhi-Firuzabad, M. Shahidehpour, Optimal demand response aggregation in wholesale electricity markets. IEEE Trans. Smart Grid **4**(4), 1957–1965 (2013)
22. S.Ø. Ottesen, A. Tomasgard, S.E. Fleten, Prosumer bidding and scheduling in electricity markets. Energy **94**, 828–843 (2016)
23. M. Vasirani, S. Ossowski, Smart consumer load balancing: State of the art and an empirical evaluation in the Spanish electricity market. Artif. Intell. Rev. **39**(1), 81–95 (2013)
24. J. Camargo, F. Spiessens, C. Hermans, A network flow model for price-responsive control of deferrable load profiles. Energies **11**(3), 613 (2018)
25. N. Ruiz, B. Claessens, J. Jimeno, J.A. López, D. Six, Residential load forecasting under a demand response program based on economic incentives. Int. Trans. Electr. Energy Syst. **25**(8), 1436–1451 (2015)
26. S. Iacovella, F. Ruelens, P. Vingerhoets, B. Claessens, G. Deconinck, Cluster control of heterogeneous thermostatically controlled loads using tracer devices. IEEE Trans. Smart Grid **8**(2), 528–536 (2017)
27. J. Saez-Gallego, J.M. Morales, M. Zugno, H. Madsen, A data-driven bidding model for a cluster of price-responsive consumers of electricity. IEEE Trans. Power Syst. **31**(6), 5001–5011 (2016)
28. O. Corradi, H. Ochsenfeld, H. Madsen, P. Pinson, Controlling electricity consumption by forecasting its response to varying prices. IEEE Trans. Power Syst. **28**(1), 421–429 (2013)
29. J. Saez-Gallego, J.M. Morales, Short-term forecasting of price-responsive loads using inverse optimization. IEEE Trans. Smart Grid **9**(5), 4805–4814 (2018)
30. M. Gonzalez Vaya, Optimizing the Electricity Demand of Electric Vehicles: Creating Value Through Flexibility, PhD thesis, ETH Zurich (2015)
31. G. Petretto et al., Representative distribution network models for assessing the role of active distribution systems in bulk ancillary services markets. 2016 Power Systems Computation Conference (PSCC), Genoa (2016), pp. 1–7
32. R. Carmona, M. Coulon, D. Schwarz, Electricity price modeling and asset valuation: A multi-fuel structural approach. Math. Financ. Econ. **7**(2), 167–202 (2013)

33. D. Pudjianto, C. Ramsay, G. Strbac, Virtual power plant and system integration of distributed energy resources. IET Renew. Power Gener. **1**(1), 10–16 (2007)
34. C. Kieny, B. Berseneff, N. Hadjsaid, Y. Besanger, J. Maire, On the concept and the interest of virtual power plant: Some results from the European project Fenix. IEEE Power & Energy Society general meeting, Calgary, AB (2009), pp. 1–6
35. P. Koponen et al., Toolbox for aggregator of flexible demand. 2012 IEEE International Energy Conference and Exhibition (ENERGYCON), Florence (2012), pp. 623–628
36. APEX-Spot, Intraday market with delivery on the French TSO zone. [Online]. Available: http://www.epexspot.com/en/product-info/intradaycontinuous. Accessed 6 Mar 2018
37. Gestor Mercati Energetici (GME), Spot electricity market in Italy. [Online]. Available: http://www.mercatoelettrico.org/en/mercati/mercatoelettrico/mpe.aspx. Accessed 6 Mar 2018
38. OMIE, Intraday market in Iberian Peninsular (Spain and Portugal). [Online]. Available: http://www.omie.es/en/home/markets-and-products/electricity-market/our-electricity-markets/intra-day-market. Accessed 6 Mar 2018
39. Hungarian Power Exchange Intraday Market Workshop (Nov 2016). [Online]. Available: https://docplayer.net/35666822-Hupx-intraday-market-workshop.html. Accessed 26 Sept 2018
40. Polish Power Exchange (POLPX) indices for the Polish Day-Ahead and Intraday Market. [Online]. Available: https://wyniki.tge.pl/en/wyniki/euroindex/description/. Accessed 6 Mar 2018
41. SmartNet, Aggregation models. Deliverable D2.1 (May 2018). [Online]. Available: http://smartnet-project.eu/wp-content/uploads/2018/05/D2.1_20180524_V1.0.pdf. Accessed 28 Feb 2019
42. SmartNet, Network and market models. Deliverable D2.2 (Feb 2019). [Online]. Available: http://smartnet-project.eu/wp-content/uploads/2019/02/2019215113154_D2.2_20190215_V1.0.pdf. Accessed 28 Feb 2019

Chapter 4
ICT Requirements in a Smart Grid Environment

Seppo Horsmanheimo, Lotta Tuomimäki, Raúl Rodríguez Sanchez, Filip Pröstl Andrén, and Claus Amtrup Andersen

4.1 Introduction

Energy systems are moving from static and centralised architectures towards more flexible and distributed structures as the share of distributed energy resources (DERs) gets bigger [6]. At the same time, the utilisation of ICT expands throughout geographically distributed networks allowing an increased number of remotely monitored and controlled subsystems and components. In particular, large-scale deployment of DERs, smart distribution systems, and real-time market will boost the utilisation of automation and ICT.

As the energy flow becomes more and more bidirectional, the amount of exchanged information increases and demands for communications become more stringent and more versatile because of the diversity of, e.g. offered ancillary services, end-users' preferences, and market regulations. For example, improved communication solutions are needed to increase controllability and to ensure high quality and flexibility of distributed energy systems. This poses challenges for today's communication systems, because the communication cost, flexibility, quality of service, availability, response time, and security do not always meet all of the expectations.

S. Horsmanheimo (✉) · L. Tuomimäki
VTT, Espoo, Finland
e-mail: seppo.horsmanheimo@vtt.fi

R. R. Sanchez
TECNALIA – Parque Tecnologico de Gipuzkoa, Donostia-San Sebastian, Spain

F. P. Andrén
AIT, Vienna, Austria

C. A. Andersen
EURISCO, Odense, Denmark

G. Migliavacca (ed.), *TSO-DSO Interactions and Ancillary Services in Electricity Transmission and Distribution Networks*,
https://doi.org/10.1007/978-3-030-29203-4_4

We also need to understand that it will take time before the new TSO-DSO coordination schemes and market models presented in Chap. 1 are realised. Meanwhile, ICT solutions will evolve making it difficult to predict the most optimal and cost-efficient communication technologies for different parts of the energy system.

4.1.1 Conceptual Reference Model

In order to select appropriate technologies, we first need to identify relevant ICT requirements in different layers of the energy system. A conceptual model depicted in Fig. 4.1 can be used for the dialogue between ICT and energy personnel to capture the main data exchange operations and their requirements in different TSO-DSO coordination schemes.

The model presents actors, system components, and services. In the figure, grey rounded boxes present core business actors/roles in different coordination schemes. The stakeholders can play multiple business actor roles. For example, an aggregator can do both technical aggregation and energy trading. The market operator (MO) role can be played by various stakeholders depending on the market scenario: central TSO (market), DSO (local), TSO-DSO (shared), or IMO (independent).

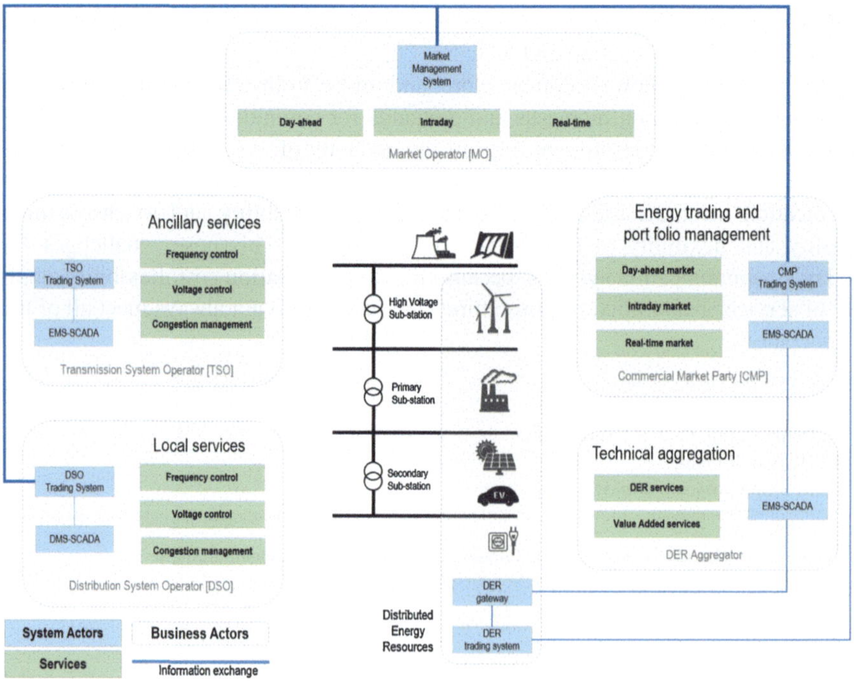

Fig. 4.1 A conceptual reference model illustrating actors, system components, and services in the energy system [19]

Blue boxes represent the main system components a.k.a. system actors or entities used by business actors. Trading system (TS) is devoted to exchange information with the market management system, e.g. schedules for prequalification, procurement, or activation of ancillary services. DMS/EMS-SCADA is considered here as the system used for network monitoring and control operations. Respectively, market management system (MMS) is dedicated for running market processes (by the TSO, DSO, or IMO) and to establish a link between the market operator and stakeholders.

Connecting blue lines represent external data exchange links between system components. Thinner lines in the figure are presenting internal communication links. We are analysing the system operations mainly from an energy market point of view. Market operations are not time critical, so they do not require an immediate response. Only DMS/EMS-SCADA-related remote control and monitoring—involving the real-time exchange of measured data—is considered time sensitive.

Green boxes represent core ancillary services including frequency and voltage control and congestion management. The pictures in the middle represent the grid infrastructure and distributed energy resources, from high voltage down to low voltage, which help mapping the energy market events to the physical grid entities.

4.2 Review of Enabling Technologies

We need to have understanding about the capabilities of the available technologies with respect to today's and tomorrow's requirements of ancillary service (AS) applications. Some ASs may have very relaxed criteria whereas others may have stringent requirements for reliability and response time. Those types of services are used mainly for control and protection of critical parts in the energy system.

Communication technologies can be divided into wired and wireless ones. The former includes communication technologies that utilise, e.g. copper and fibre-optic cables that can be considered as reliable and secure communication technologies. On the other hand, wireless technologies are more flexible, agile, and less expensive to deploy and operate. They are more compelling for utilities and aggregators as communication network performance and security have improved significantly and end devices have become less expensive. Unfortunately, there is no one-size-one-technology-fits-all solution available.

Next, we go through some potential wireless technologies introduced for machine to machine (M2M) and Internet of things (IoT).

4.2.1 Existing Mobile Communication Technologies

Mobile communication technologies allow us to communicate with others in different locations without the use of cables. These technologies exploit cellular network infrastructure distributed over a wide geographical area. Each cell has a base station in a fixed location. Cells together provide wide radio coverage so that a mobile terminal can communicate while moving through cells during the data transmission. The high availability, low operational and device costs, and steadily increased performance and quality have made those technologies very compelling for energy systems.

GSM (Global System for Mobile Communications), UMTS (Universal Mobile Telecommunications Service), and LTE (Long-Term Evolution) are examples of worldwide licenced mobile communication technologies.

GSM is a second-generation (2G) cellular technology offering digital voice calls and limited data services (SMS and MMS messages). GSM has evolved significantly over the years. General Packet Radio Service (GPRS) added packet-switched functionality to GSM for always-on data connection. It also offered higher data rates by aggregating several GSM time slots. The speed is around 14 kbit/s in the uplink direction (from a terminal to a base station) and 40 kbit/s in the downlink (from a base station to a terminal). EDGE (Enhanced Data rates for Global Evolution) technology improved data rates to few hundreds of kbit/s. The EDGE is generally seen as a 2.5G technology between GSM and UMTS.

UTMS is the third-generation (3G) cellular technology that offered greater spectral efficiency and higher bandwidths enabling multimedia services and Internet access. UMTS has evolved over the years. High-speed packet access (HSPA) and evolved high-speed packet access (HSPA+) are offering over 10 Mbps data rates. They are using improved high-speed download packet access (HSDPA) and an uplink equivalent (HSUPA) protocols. This allows the multimedia services, interactive gaming, and large file downloads.

LTE is the most recently deployed technology. It is a fourth-generation (4G) mobile technology offering high-speed data services and more flexible use of frequency allocations. LTE base stations are collaborating directly and they can exchange information about, e.g. current load levels, predicted capacity, and coverage levels in order to steer their coverage and capacity according to the cellular network load. For example, if demand increases at a particular location, a base station can shrink its coverage in order to increase capacity locally. The excluded areas are then serviced by other base stations. Decisions made in base stations flatten the LTE architecture and improve the cellular network performance and reliability. LTE exploits also advanced antenna techniques and beamforming to increase data rates and reliability. Improvements in data rates in 3G and 4G mobile technologies are presented in Fig. 4.2.

One complementary mobile communication technology designed for critical communication especially for authorities is TETRA (Terrestrial Trunked Radio). This technology is also used by several DSOs and TSOs. TETRA is designed to provide secure, reliable, and robust communication services. It offers Short Data Service (SDS), which is comparable with GSM's Short Data Message (SMS) over a guaranteed and secure pipe. One of the drawbacks of TETRA technology is that it is deployed by a relatively small end-user group, so service and device costs are rather high compared to other mobile technologies. TETRA is a narrowband technology like GSM having very limited data transmission capabilities.

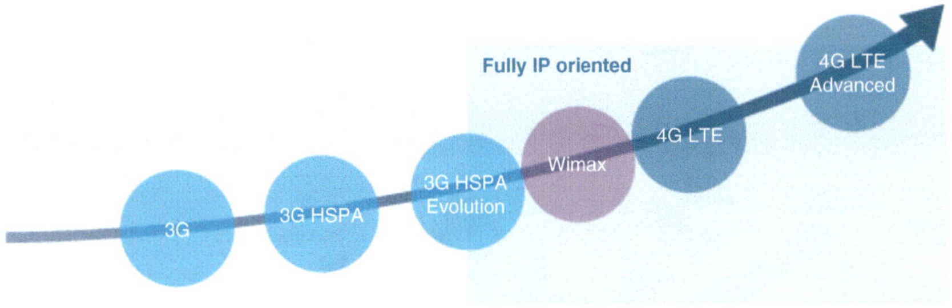

Market impact	2003	2006	2009	2010	2016
Peak rate	384 kbps	7 Mbps	42 Mbps	~150 Mbps	~1 Gbps
Typical user rate downlink	~200 kbps	1–2 Mbps	2–10 Mbps	10–20 Mbps	~30–100 Mbps
Typical user rate uplink	64 kbps	64–884 Mbps	0.5–4.5 Mbps	5–10 Mbps	~10–60 Mbps

Wifi for supporting hot spot coverage

Fig. 4.2 Evolution of cellular technologies and data rates [15]

The current trend is towards converging LTE and mission critical TETRA technologies. Three evolution paths are proposed: one looking at LTE (and its future releases or 5G) as a replacement, and another seeing LTE as a complementary technology to be used together with TETRA to give the best of both technologies (for example, voice over TETRA and data over LTE). The third option is to adopt a range of different standards for both narrow and broadband technologies according to their operational and business requirements, spectrum availability, and legislation.

4.2.2 Future Mobile Communication Technology

The next mobile generation is 5G which is no longer intended exclusively for human communications. The volume of data and information communicated with and between things and machines is anticipated to increase drastically. Moreover, applications will span from traditional voice and video applications towards industrial automation, virtual reality, automated driving, and robot applications as illustrated in Fig. 4.3.

There are three main performance dimensions defined for 5G:

- *Massive machine-type communications (mMTC)* is designed to connect millions of inexpensive sensors and machines. Contention-based and connectionless access procedures are supported. The connectivity is evaluated in terms of how well the applications work regardless of device type, time, or location.

Fig. 4.3 5G verticals and potential applications [9]

- *Ultra-reliable and low latency communication (URLLC)* is designed to connect more complex devices having stricter requirements on reliability and availability. While data transmissions from mMTC devices are infrequent and not delay dependent, ultra-reliable machine-type communication (uMTC) addresses services with high reliability and short latencies. uMTC services are typically safety or mission critical.
- *Enhanced mobile broadband (eMBB)* is designed to provide capacity enhancements. It is considered as extended support of conventional mobile broadband through improved peak, average, cell-edge data rates, capacity, and coverage.

Figure 4.4 presents the main 5G key performance indicators (KPIs) characterising 5G dimensions with different colours: eMBB (green), URLLC (yellow), and mMTC (blue) [12].

5G is planned to be launched around the year 2020 according to the time schedule in Fig. 4.5. It is designed for carrying mission critical, massive machine type of traffic and connecting efficiently other types of devices than mobile terminals.

The standardisation of 5G is still an ongoing process, but some key 5G requirements and capabilities have been agreed.

5G will be a single radio access network (single RAN) technology that is built upon new radio access technologies and evolved existing ones like LTE, HSPA, GSM, and Wi-Fi. The benefit of the single RAN technology is that mobile operators can simplify their cellular network architecture by operating different radio technologies on a single multipurpose hardware platform. This platform will exploit both licensed and unlicensed bands. The latter ones are used to provide additional capacity in the best effort manner.

5G will be a unified and programmable infrastructure that offers a scalable service experience everywhere and anytime. This means that changes in logical cellular network architecture can be done simply by software updates. This flexibility

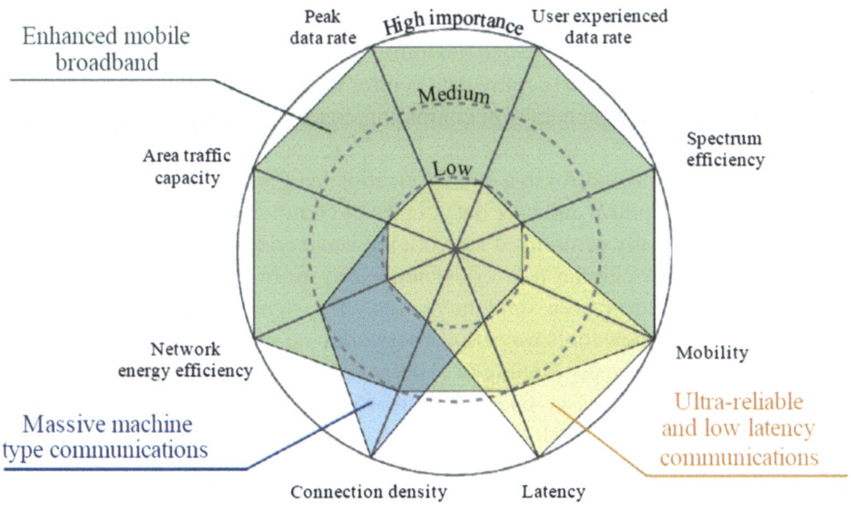

Fig. 4.4 5G KPIs according to service type [12]

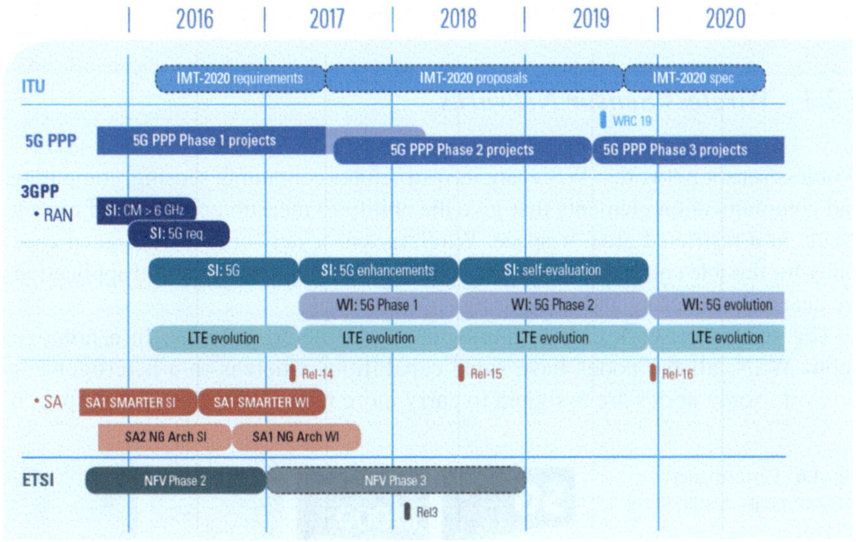

Fig. 4.5 Standardisation timeline [3]

will be achieved by exploiting upcoming architectural trends like network clouds, software-defined networking (SDN), network function virtualisation (NFV), multi-access edge computing (MEC), and fog computing (FC).

Network clouds allow resource pooling that reduces overprovisioning and underutilisation of cellular network resources. For example, some base stations dedicated for low-latency services could be connected directly to a small nearby data centre, whereas in the case of no latency critical services, the connection can be to a large data centre further away. Such flexibility would allow mobile network operators or energy stakeholders to deploy data centres of different sizes to meet specific service needs. The more general fog computing concept offers an architecture that exploits end-user clients or near-user edge devices to carry out storage, communication, computation, and control in a communication network.

SDN decouples control and user data planes of communication network devices and provides a logically centralised network view and control. In a complementary way, NFV decouples cellular network functionality from dedicated hardware and promotes the software-driven implementation of system functionality. Together SDN and NFV offer new tools for cellular network load optimisation and improved resilience. They enable repositioning of cellular network functions according to network load, service quality, or operational reasons. Multi-access edge computing will offer a solution to move cloud computing capabilities and an IT services at the edge of the mobile network, which enables higher reliability and lower latency for real-time data exchange.

The main enhancements of 5G with target numbers are presented in Fig. 4.6.

Table 4.1 summarises main 5G trends focusing on reliable, high-quality, and flexible data services.

4.2.3 *Wireless Sensor Networks*

Wireless sensor networks (WSN) are infrastructures containing sensing, computing, and communication elements that give the ability to measure, collect, and react to events in a restricted area or space. Wireless sensor networks are designed especially for flexible communication. Typical wireless sensor network grid applications are near-field metering and monitoring applications.

The sensor network can be homogeneous or heterogeneous. In a homogeneous WSN, all the nodes have same capabilities whereas in a heterogeneous network, some nodes are assigned to carry more responsibilities with respect to

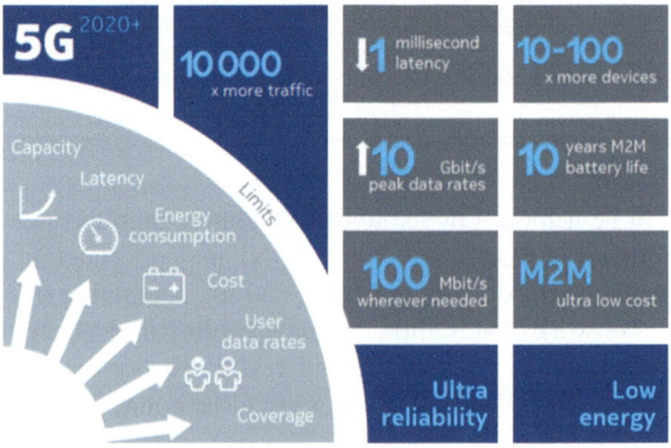

Fig. 4.6 Performance enhancements designed for 5G [13]

Table 4.1 A summary of major trends in 5G

Trend	Description
Low latency and high reliability	Several scenarios require very low latency and very high reliability. For instance, in the case of industrial control, where the 5G system is deployed to enable wireless remote control of industrial devices to accomplish complicated and precise time-sensitive actions, the expected latency should be no longer than 1 ms while the expected reliability being 99.999%.
Network slicing	Network slicing allows the operator to provide deeply customised networks. Different requirements on functionality (e.g. charging, policy control, security) and performance (e.g. latency, mobility, and throughput) can, therefore, be realised with the same physical networks by offering dedicated network slices for different end-users.
Multiple access technologies	5G will support 3GPP access technologies, including one or more 5G RATs and E-UTRA as well as non-3GPP access technologies, to ensure interoperability. For optimisation and resource efficiency, the 5G system will select the most appropriate 3GPP or non-3GPP access technology for a service, potentially allowing multiple access technologies to be used simultaneously for one or more services on a device.
Priority, QoS, and policy control	As 5G network will support many commercial services, and some of them require priority treatment, the network will offer a means to provide the required QoS, e.g. guaranteed bandwidth, low latency, and high reliability, and to prioritise resources. While existing QoS and policy frameworks handle latency and improve reliability by traffic engineering, it is necessary for the 5G network to offer QoS and policy control for reliable communication with latency required for a service and enable the resource adaptations as necessary. 5G network QoS aims for an E2E covering not only RAN and core network but also backhaul and network-to-network interconnections.
Connectivity models	In a 5G network, the UE can connect to the network directly or connect with the network using another UE as a relay UE, or they may be capable of using both types of connections in multi-hop mode.
Network capability exposure	Future UEs are equipped with various types of sensors (e.g. accelerometer, gyroscope, magnetometer, barometer, proximity sensor, GPS) and have access to various radio access networks. 5G network will exploit this information with data from supported access technologies, application context, and traffic characteristics to optimise radio resource utilisation.
Self-backhaul	The 5G network will support the increasingly high densification of access nodes with wireless backhaul, which enables simpler deployment and incremental rollout. Network planning and installation efforts can be reduced by leveraging plug-and-play-type features including self-configuration, self-organising, and self-optimisation.
Extreme long range coverage in *low-density* areas	As a fully connected society is expected in the near future, 5G network will enable access everywhere over long distances (e.g. in extremely rural areas or at sea) including both human and machine users. To realise the above explained capability, one significant difference in 5G design principles is a layered architecture with separated control and user planes. In this way, the control plane enables microservices architecture through single and modular RAN-CN interface, supports network slicing, and controls APIs for mobile edge computing, while the user plane focuses on providing connectivity function with rigorous requirements on latency, throughput, etc. Furthermore, the separated control and user planes can also be easily scaled up or down, respectively, to meet specific user needs.

the sensor network load and control. WSN applications can be divided into time-driven, event-driven, and query-driven. The classification is done based on the sensor network activity.

In a time-driven implementation, sensors will transmit their readings periodically. Sampling and communication occur periodically meaning that the communication times are known beforehand. In an event-driven sensor network, the sensors monitor the area and transmit information only when something meaningful happens. The attempt is to minimise the data traffic and transmission of redundant information. The last category is query-driven systems where gathered information is stored locally in the sensor nodes and required information is retrieved with queries. Scheduled communication protocols are typically used in time-driven and on-demand protocols in event-driven implementations.

One of the widely deployed sensor network topology is a cluster star. It is a hybrid topology formed from star and mesh network topologies. Sensor nodes are divided into normal nodes and cluster heads a.k.a. sinks. Normal nodes are only communicating with their cluster head whereas cluster heads can also communicate with each other. This flexible structure allows most of the sensors in the network to be very simple, and only a few nodes need to have additional memory and processing capabilities. The communication link between the sensor network and global network is typically through dedicated gateways using licensed or unlicensed radio access technologies.

4.2.4 Low-Power Wide Area Networks

Low-power wide area networks (LPWAN) are designed for M2M applications that have a long communication link (even a few hundred kilometers), low data rates, and long battery lives. They can operate unattended for long periods of time and at frequencies below 1 GHz, because it offers longer range and better building penetration, e.g. in indoor spaces.

LPWA technologies are split into two subcategories (see Table 4.2). The current proprietary LPWA technologies, such as SigFox, LoRa, M-Bus, or Dash-7, operate in unlicensed spectrum, while Clean Slate and 3GPP-standardised cellular IoT technologies, e.g. NB-IoT and NB-LTE-M, operate in licensed spectrum. LTE-M and NB-LTE-M are supplementary solutions addressing different use cases. LTE-M has higher capacity but NB-LTE-M has slightly lower cost and better coverage. Differences between different LPWA technologies are presented in Table 4.2.

4.2.5 Unlicensed LPWA Technologies

SigFox, LoRa, Wireless M-Bus, and Dash-7 are the best known unlicensed band LPWA technologies.

Table 4.2 LPWA IoT connectivity overview [14]

	SigFox	LoRa	Clean Slate	NB LTE-M Rel. 13	LTE-M Rel. 12/13	EC-GSM Rel. 13	5G (targets)
	sigfox	LoRa	cIoT	lte	lte	GSM	5G
Range (outdoors)	<13 km	<11 km	<15 km	<15 km	<11 km	<15 km	<15 km
MCL	160 dB	157 dB	164 dB	164 dB	156 dB	164 dB	164 dB
Spectrum bandwidth	Unlicensed 900 MHz 100 Hz	Unlicensed 900 MHz <500 kHz	Licensed 7–900 MHz 200 kHz or dedicated	Licensed 7–900 MHz 200 kHz or shared	Licensed 7–900 MHz 1.4 MHz or shared	Licensed 8–900 MHz 2.4 MHz or shared	Licensed 7–900 MHz shared
Data rate	<100 bps	<10 kbps	<50 kbps	<150 kbps	<1 Mbps	<10 kbps	<1 Mbps
Battery life	>10 years	>10 years	>10 years	>10 years	>10 years	>10 years	>10 years
Availability	Today	Today	2016	2016	2016	2016	Beyond 2020

SigFox is a narrowband (or ultra-narrowband) technology with low noise level. It is bidirectional, but its capacity to downlink direction (i.e. from the base station to the endpoint) is more limited. SigFox owns all of its technology from the backend data and cloud server to the endpoint software, but it has opened its endpoint technology to silicon manufacturers and vendors. The business idea is to allow the applications to be very inexpensive and offer already-installed nationwide networks. The drawback is that only one SigFox network can be deployed in an area due to exclusive arrangements with the selected network operator. Moreover, the technology is not applicable for continuous communication due to the relatively high latency with low predictability.

LoRa is a wideband CDMA technology with an inherently higher noise level. Due to efficient coding, communication link budget figures are about the same as in SigFox. LoRa uses the same radio on a base station and at endpoints. Consequently, the cost of a LoRa terminal is higher than a SigFox terminal but a LoRa base station is cheaper making the overall technology less expensive for network deployment. LoRa ecosystem is open, so anyone, e.g. large network operators, private companies, and start-ups, can basically build and manage their own networks. However, there are open issues related to the roaming from public to public and from private to private network. Although the LoRa ecosystem itself is open, it contains a black box element. Semtech is the only company that makes the radio for LoRa.

Wireless M-Bus has been specifically standardised for the smart grid domain. The interface is M-Bus, and the wireless part is merely an extension. Although Wireless M-Bus has been deployed for advanced metering system (AMS), the limitation to IPv4 may restrict its deployments in Asia [10].

The last unlicensed LPWA technology is Dash-7, which originates from the ISO/IEC 18000-7 standard. The technology was used for military logistics, but has evolved to support mid-range LPWA applications. The network topology is a tree or star. The technology forces end devices to check the channel periodically for possible downlink transmissions. As a result, Dash-7 has much lower latency for downlink communication than other LPWA technologies but at an expense of higher energy consumption [16].

Table 4.3 summarises the key features of the four unlicensed LP-WAN technologies.

4.2.6 Licensed LPWA Technologies

The challenge with unlicensed systems is that the communication is not guaranteed and other devices can use the same frequency band and interfere with the communication. 3GPP introduced MTC (machine-type communication) in LTE to cover machine-to-machine communications including all types of data communication without human intervention. In addition to the conventional GSM and LTE, three new technologies—eMTC, NB-IoT, and EC-GSM-IoT—for licensed cellular IoT technologies have been standardised in 3GPP [1, 2]. Their features are presented in Table 4.4.

Table 4.3 Properties of unlicensed LP-WAN technologies

	SigFox	LoRa	M-Bus	Dash-7
Frequency (MHz)	865-868/ 902-928	EU: 433/868 USA: 780/915	868, 169	EU: 433/868 USA: 780/915
Channel width (Hz)	100 Hz	≥125 kHz	10 kHz to 100 kHz	25 or 200 kHz
Transmitted power (dBm)	Up to 20	EU: 14 dBm USA: 27 dBm	10 dBm	433 MHz: 10 dBm 868/915 MHz: 27 dBm
Topology	Star	Star	Star	Star, tree, mesh
Uplink data rate	4 × 8 b/day	EU: 30 b/s–50 kb/s USA: 100–900 kb/s	4.8–100 kb/s	9.6–167 kb/s
Downlink data rate	100 b/s	EU: 30 b/s–50 kb/s USA: 100–900 kb/s	4.8–100 kb/s	9.6–167 kb/s
Battery life	10 years	10 years	Years	10 years
Support for IPv6	Unlikely	Likely	No	Likely
Governing body	SigFox	LoRa alliance	M-Bus	Dash-7 alliance
Deployment status	Deployed since 2009	Deployed	Available	Deployed since 2015
Nodes per gateway	1,000,000	250,000	Not specified	N/A (connectionless)
Est. costs ($)	Node: 2	Node: 30	Node: 10	Node: 2

Table 4.4 Licenced MTC and IoT technologies

	GSM (Rel. 8)	EC-GSM-IoT (Rel. 13)	LTE (Rel. 8)	eMTC (Rel. 13)	NB-IoT (Rel. 13)
LTE user equipment category	N/A	N/A	Cat. 1	Cat. M1	Cat. NB1
Range	<35 km	<35 km	<100 km	<100 km	<35 km
Max. coupling loss	144 dB	164 dB	144 dB	156 dB	164 dB
Spectrum	Licensed GSM bands	Licensed GSM bands	Licensed LTE bands in-band	Licensed LTE bands in-band	Licensed LTE in-band guardian band Stand-alone
Bandwidth	200 kHz	200 kHz	LTE carrier bandwidth (1.4–20 MHz)	1.08 MHz (1.4 MHz carrier bandwidth)	180 kHz (200 kHz carrier bandwidth)
Max. data rate[a]	<500 kbps (DL/UL)	<140 kbps (DL/UL)	<10 Mbps (DL) <5 Mbps (UL)	<1 Mbps (DL/UL)	<170 kbps (DL) <250 kbps (UL)

[a]Max. data rates provided are instantaneous peak rates

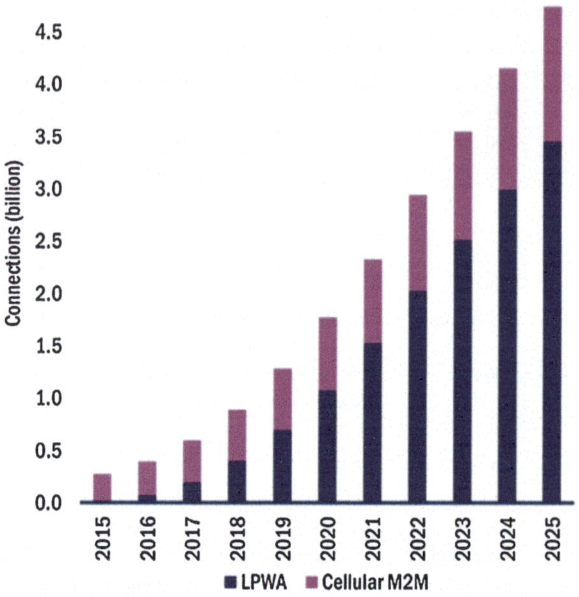

Fig. 4.7 Analysys Mason's forecasts for LPWA and cellular M2M connections 2015–2025 [4]

The specification of NB-IoT builds on synergies of existing mobile network infrastructure. It provides an extension to LTE with flexible deployment options. The 3GPP specifications allow three NB-IoT deployment options: stand-alone, guard band, and in-band.

LPWANs are an emerging area of the IoT and they represent a huge market opportunity as the IoT matures. This technology is anticipated to create new M2M use cases where connectivity costs are expressed in a few dollars per year rather than per month. It is estimated that there will be 3.5 billion LPWA connections and 1.3 billion cellular network connections by the year 2025 [4]. That is equivalent to the current number of global cellular subscriptions, but the density of connected devices is likely to be less uniform (Fig. 4.7).

The deployment of IoT for energy systems will gradually extend towards utility and transport networks by 2020. The goal after 2025 is to have plug-and-play smart objects, which can be installed in any environment with an interoperable backbone allowing them to communicate with other smart objects in the vicinity.

4.2.7 Service Architectures

Future energy systems do not only struggle with how to coordinate operation of millions of different devices and subsystems but also how to manage and orchestrate an increasing number of services. The bad performance and inflexibility may be result from the inadequate service architecture. IT enterprises have been struggling with constantly changing application versions and upgrades, integration, and security issues, etc., so different types of service-oriented architectures (SOAs) have been designed to cope with challenges.

4.2.7.1 Enterprise Service Bus

The core concept of the enterprise service bus (ESB) architecture is that different applications are integrated by putting a communication bus between them and enabling each application to talk via the bus. This decouples applications and services from each other allowing them to communicate without dependency or knowledge of other applications on the bus. ESB provides basic adoption, translation, routing, and so-called commodity services including event handling and queuing, data transformation and protocol conversions, security, and exception handling. ESB defines a set of rules and principles for integrating numerous applications together.

ESB architecture is suitable for large monolithic systems with in-built security and service orchestration. Therefore, it is applicable in TSO-DSO level and could be bought as a turnkey solution from big service providers. However, ESB solutions are too expensive and too inflexible at the aggregator and DER levels where heterogeneity of applications and small mobile or embedded devices is higher. Moreover, applications and services have evolved since ESB was introduced. Nowadays, services are more distributed with significantly shorter lifecycles, e.g. mobile and IoT applications. The conventional all-or-nothing ESB implementation is too expensive and restrictive for smaller enterprises, so new complementary architecture concepts were needed.

Integration of monolithic and agile IoT services is a challenging task. To lower the complexity, complementary technologies such as SOA Gateways and microservices are applied. They are simpler and thus can enable a lightweight deployment with higher agility and lower cost.

4.2.7.2 SOA Gateway

SOA Gateway was initially created to protect internal applications when interfaces protected only by firewalls are being exposed to external parties. SOA Gateway offers additional security and is applicable for dividing the secure centralised core system parts from the distributed and less reliable system parts.

The gateway is typically deployed as a hardware component that seamlessly controls access to services, protects information through data-level encryption, ensures the integrity of a message through signatures, and controls corporate information flow.

Key benefits of SOA Gateway are as follows:

- Scales from point solutions to enterprise-wide deployment
- More configuration rather than integration
- No central rules or brokers

- Easy to plug in and plug out and loosely coupling system
- Incremental upgrades and patches without service interruptions

On the other hand, key disadvantages are as follows:

- Slower communication speed compared to compatible services.
- Single point of failure can bring down all communications in the enterprise.
- High configuration and maintenance complexity.

The strengths of SOA Gateway are related to security, high-performance transformation, and edge-based protocol mediation. SOA Gateway can act as a bridge between different technologies, which is important in order to build integrated enterprise and IoT/mobile services. It also offers secure extensions to applications deployed in public or private cloud environments.

4.2.7.3 Microservice Architecture

In microservice architecture, an application or service is decomposed into multiple small, granular, independently deployable services. Microservices are designed with agility in mind [7]. Services are very simple and they focus on doing only one function well. As a result, they are easier to test and validate, which ensures higher service quality.

The architecture encourages developers to implement IT solutions as microservices without using any intermediate integration products such as ESB. Since the parts are independently deployed, they can also be independently scaled to fulfil different end-user needs.

Microservice architecture offers the highest flexibility and support for IoT and mobile services with low OPEX and CAPEX costs, because:

- Each service can be built with the best-suited technologies and tools, allowing high flexibility for implementation.
- Multiple software developers can deliver services independently enabling continuous delivery and frequent releases while keeping the rest of the system stable.
- In case a service goes down, it will only affect the parts that directly depend on the service. The other parts will continue to function well.

In practice, microservices cannot be used alone in energy systems, since large TSOs, DSOs, and aggregators have internal proprietary or legacy systems, which cannot be converted into microservices. Instead of focusing on one service architecture, the most suitable and flexible solution for future energy system would be to take advantage of all of them. Microservices can be used to address specific service cases executed at the edges of the grid. ESB can take care of service orchestration in core grid and cater all integration needs requiring high security and reliability. SOA Gateway can act as a bridge between ESB and microservice environments.

4.2.8 Data Hub

Data hubs are recently deployed also in energy systems. Data hub is a central service platform that facilitates transparent and neutral exchange of market information and execution of business processes between all market parties [21].

Metering operators are responsible for collecting metered values directly from smart meters. The metered values are sent to a data hub through standardised processes, timeframes, and communication formats. A data hub contains the data necessary for consumption settlement and execution of market processes, such as master data related to consumers and metering points, metered values with a relevant time resolution, and historical data for analysis purposes. In a data hub, the transactions and actual metered values are interlinked with the identified metering point and can be traced upwards or downwards in the executed processes.

Additionally, a data hub provides other functionalities such as calculations of settlement and imbalance data as well as providing data for reconciliations and aggregation processes by the market parties. Information about how consumers and market players are related is managed through standardised market processes. Figure 4.8 illustrates the difference between traditional communication between suppliers and DSOs before and after data hub deployment. Dedicated point-to-point connections are replaced with a common data hub connection.

The benefits of having data hubs is a single set of services that all the system actors can connect to and trust rather than having many peer-to-peer connections and also to have a coherent and verified data set in a well-specified format.

One drawback is the possibility of a single point of failure in the system. However, a data hub is seldom hosted on a single server. It is typically distributed, even geographically, in a dedicated server farms and hosted by certified data management companies. This means that the overall reliability and availability of the system is higher or at least in the same level than having peer-to-peer networks of servers at different companies' premises.

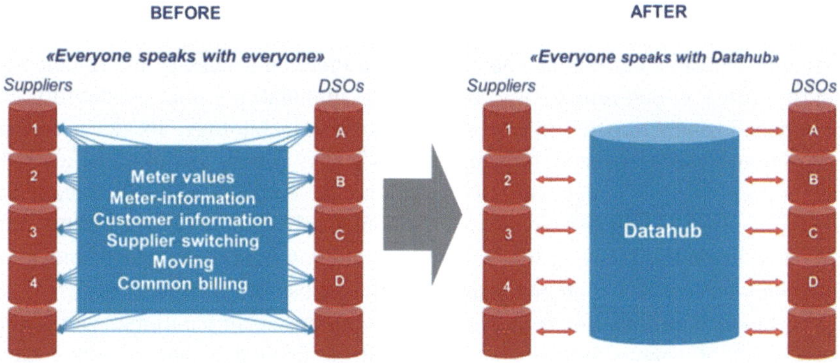

Fig. 4.8 Communication before and after a data hub deployment

4.2.9 Blockchain

Blockchain is the technology that was invented to create the peer-to-peer digital cash Bitcoin in 2008. Blockchain can be regarded as an electronically distributed ledger that is essentially an asset transaction database that can be shared across a network of multiple sites, geographies, or institutions. All participants within a network can have their own identical copy of the ledger. Any changes to the ledger are reflected in all copies in minutes or, in some cases, seconds. The assets can be financial, legal, physical, or electronic. The security and accuracy of the assets stored in the ledger are maintained cryptographically through the use of keys and signatures. Entries can also be updated by one, some, or all of the participants according to rules agreed by the network [22].

The uniqueness of this technology lies in the fact that blockchains are maintained by a shared or distributed network of participants—not by a centralised entity. This means that there is no central validation system. By design, a blockchain is inherently resistant to modification of the data.

Transactions can be created collaboratively by multiple writers, without either party exposing themselves to security threats. This is what allows delivery versus payment settlement to be performed safely over a blockchain without requiring a trusted intermediary. Another important feature of distributed ledger technologies (DLT) is the extensive use of cryptography to store, secure, and validate asset transactions.

The blockchain is, in essence, an open permissionless system where all participants can contribute to the validation process. However, the use of blockchains or distributed ledger technologies in energy trading markets could be a permission-based system with authorised participants only.

Digital cryptocurrencies like Bitcoin was implemented with blockchain 1.0. Since then, blockchain technology has evolved. Blockchain 2.0 provides smart contracts that are more extensive than simple cash transactions. Blockchain 3.0 extends the application domain beyond currency, finance, and markets covering domains like health, government, science, literacy, culture, art, and energy.

Potential uses of blockchain technology in the energy domain are quite diverse from the microenergy market to energy and flexibility trading. The blockchain technology has the potential to make trading processes far more efficient, lower the cost of trading, improve regulatory control, and eliminate unnecessary intermediaries. Security, privacy, non-repudiation, traceability, immutability, and availability are fundamental characteristics inherent to blockchain technology. Its decentralised approach and peer-to-peer architecture makes it very robust. A failure of a single node or even multiple nodes will not break down the entire system. Also, real-time processing of mass data and payment/settlement as a by-product in the trading process could be realised. Use of blockchain enforces common data formats and communication protocols, which promotes also future cross-border operations.

The blockchain technology has also shortcomings. Computational work consumes a lot of energy and how to deal with outdated data versus active data. Blockchains will always be less performant than centralised databases; lack of privacy, transaction data should not be accessible to all participants, but to authorised users only. Moreover, blockchain technology is not yet mature. It is evolving fast and a rich ecosystem of players experimenting with different blockchain variants is emerging. Apart from the Bitcoin case, proof of the technology promises in other domains has still to be delivered. It may take years before a blockchain technology is suitable for ancillary services. Although blockchain technology has a lot of potential in energy and flexibility trading, it will be economic, legal, and regulatory issues that determine whether blockchains will be used.

4.3 Analysis Process and Classification of ICT Requirements

The provision of ancillary services (ASs) from distribution networks involves the coordination between different actors and systems. Data exchanged among them contains ICT requirements that need to be known during the system design phase.

The Smart Grid Architecture Model (SGAM) created by Smart Grid Coordination Group/Reference Architecture Working Group (SG-CG/RA) [5] presents a structured approach for modelling the Smart Grid architecture. The basis for the SGAM is a three-dimensional framework consisting of domains, zones, and layers that are used to distinguish process and information managements. The physical domains present the electrical energy conversion chain including generation, transmission, distribution, distributed energy resources (DERs), and customer premises. Hierarchical zones represent the power system management entities including market, enterprise, operation, station, field, and process. Domains and zones (two-dimensional axes) form five abstract interoperability layers that are depicted in Fig. 4.9 and shortly described in Table 4.5.

The SGAM approach can also be used for modelling the proposed TSO-DSO coordination schemes with ICT requirements.

In business and function levels, existing ICT solutions are pretty ready for supporting different coordination schemes. The readiness comes from the fact that ICT infrastructures are providing services to multiple industry sectors. In addition to that, TSOs and DSOs have other more stringent communications needs, e.g. grid protection that poses requirements that exceed the ones defined by the coordination schemes. Available and future communication technologies, presented in the previous chapter, provide a good set of alternatives to fulfil the high-level communication requirements between interacting systems.

4.3.1 Data Protocols

Data protocols are required to convey information over a communication link. From ICT's viewpoint, the most relevant ancillary service procedures from the market side are prequalification and procurement. At the control side, activation and

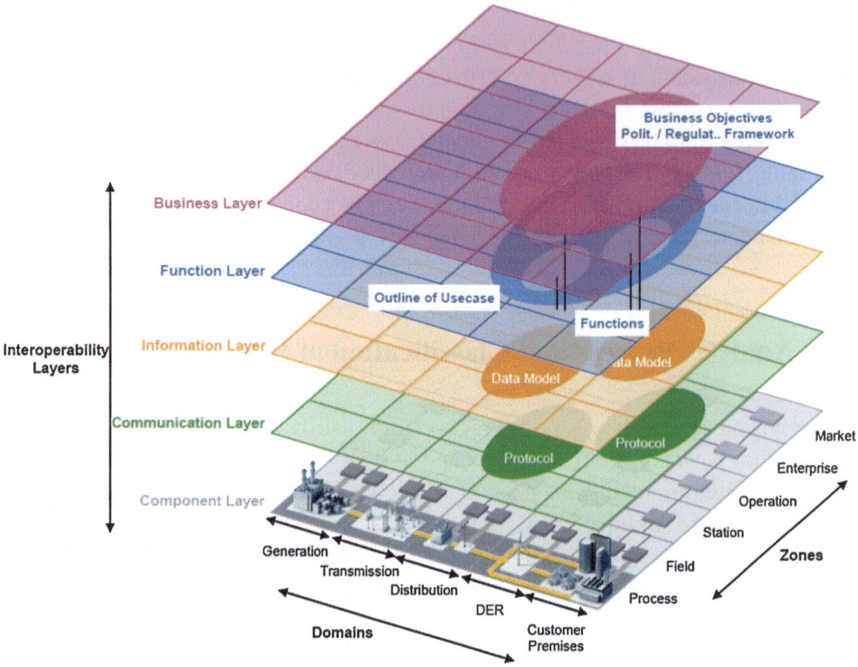

Fig. 4.9 Smart Grid Architecture Model (SGAM) [5]

Table 4.5 SGAM layers

Layer	Description
Component	It presents the physical distribution of all participating components in the smart grid system including actors, applications, network infrastructure, servers and computers, etc.
Communication	It describes protocols and mechanisms for the interoperable exchange of information between energy system components.
Information	It describes the data being used. Information objects and data models define the common semantics for functions and services to allow an interoperable information exchange.
Function	It describes functions and services and their relationships to use cases' functionality. Functions are represented independent of actors and physical implementations.
Business	It represents the business view on the information exchange. It maps regulatory, market structures, business models, and business processes of the involved market parties.

settlement are the most critical ones. The presented coordination schemes appear to be very different from market model perspective, but from the ICT's viewpoint, they have a lot of similarities. Only data exchange links between sources and targets change. In case of distributed market models, tighter coordination between TSOs and DSOs as well as between other different actors and systems is needed. Increased coordination affects especially interoperability and security requirements.

The content of information being exchanged and fulfilling associated requirements are vital to enable fluent, efficient, reliable, and secure data exchange between components in the energy systems. The amount of information sent and received and their criticalities affect the data management, security requirements, and communication and computation loads. The following data protocol properties need to be considered:

- Data structures and formats
- Data size
- Implementation complexity
- Availability and cost-efficiency
- Open or restricted (international/de facto/proprietary standard)
- Security
- Legacy issues
- Available communication technologies

Regarding to the data being exchanged, suitable protocols and standards already exist. For example, Smart Energy Grid Coordination Group (SG-CG a.k.a. SEG-CG) has provided recommendations for suitable communication and information standards to achieve interoperability in energy systems throughout Europe. The SG-CG recommends the following standards involving market interactions [11]:

- EDI: it is not really a standard but a library by ENTSO-E containing several documents and definitions for the harmonisation and implementation of standardised electronic data interchanges in the context of achieving EU energy policy goals. The Market Data Exchange Standard (MADES) is comprised of standard protocols and it utilises IT best practices to create a mechanism for exchanging data (documents) over any TCP/IP communication network and to facilitate business to business information exchanges as described in IEC 62325-351 and IEC 62325-451 standards.
- IEC 62325: it is a set of standards describing a framework for energy market communications. Its main parts are covering the communication between market participants and market operators. The common information model (CIM) specifies the basis for the semantics for this message exchange.

IEC 62325 and ENTSO-E EDI standards provide core foundations for ancillary service market processes. Table 4.6 shows a summary of exchanged data types supported by EDI and IEC 62325 standards.

IEC 62325 includes data formats for, e.g. bids, market results, acknowledgement, and settlements. IEC 62351 is considered as a reference standard for security in smart grid environments. It is aimed at improving security in automation systems in the power system domain.

The standards proposed at European level pursuing interoperability must be used whenever possible. They set high-level requirements for the design of the communication architecture. Proprietary solutions should be limited to cases where existing standards cannot be deployed. For example, an end device in a process or field zone has too low computation capacity to run complex standardised protocols. Here, the use of de facto standards is justified. If a dedicated gateway component is

Table 4.6 Market data types in ICT standards and protocols

Type of data to be exchanged	EDI	IEC 62325
Market context definition	X	X
Market document/messages exchange (secure)	X	X
Market business process implementation methodology	X	
Capacity allocation and nomination: congestion management and scheduling	X	X
Acknowledgement of business process in markets		X
Scheduling information	X	X
Reserves resources information: tendering planning and activation	X	X
Settlement data (imbalance reports, metered information, finalised schedules, etc.)	X	X
Problem settlement and status request in market processes		X
HVDC scheduling	X	
Information for interconnection capacity determination from critical network elements	X	

used to convert information between standardised and de facto protocols, then the use of de facto standard has a low impact on overall interoperability.

Security is getting increasingly more important as remote control and system automation is extended to the edges of the grid. Encryption and use of digital certificates are already widely deployed in market platforms, but this does not diminish the relevancy of considering integrity, availability, confidentiality, authentication, and non-repudiation aspects in communication. Improved security can also degrade the system performance by increasing overall latency, volumes of exchanged data, and requiring significantly more processing capacity. The system design is often a compromise of performance, cost, and security.

As a summary, existing data exchange protocols and standards offer good foundation and suit well to the proposed coordination schemes. The main question in the future will be more related to privacy and data ownership in case of cross-border systems.

4.3.2 Process of Capturing ICT Requirements

In the SmartNet project [20], a process to identify communication and ICT requirements in five TSO-DSO coordination schemes was developed. The focus was on ancillary services covering frequency control (FC), automatic and manual frequency restoration reserve (aFRR/mFRR), and voltage control (VC) presented in Chap. 1. Figure 4.10 shows in rows the five distributed and centralised market models and in columns the investigated ancillary services. All ancillary services are not relevant to all of the coordination schemes, so the excluded ones are marked with red minus symbols in Fig. 4.10. For an example, frequency control

Fig. 4.10 Mapping of ancillary services and coordination schemes [17]

is not relevant in cases of local AS, shared balancing responsibility, and integrated flexibility market models and voltage control in centralised AS or integrated flexibility market models.

For discovering ICT requirements in different TSO-DSO coordination scenarios, the SGAM framework is applied together with IEC 62559 and ELECTRA's use case design methodologies and design templates [8]. The SmartNet project extended this process by adding ICT requirements in each layer of the SGAM model. The developed process is iterative and incremental. The iterative approach is used to refine requirements and to discover possible gaps in the design by studying revised specifications. The incremental approach is used to extend the design in a step-by-step manner from the business layer down to the component layer.

The process is divided into three stages: (i) classification of ICT requirements, (ii) harmonisation of ICT requirements and creation of the common architecture model, and (iii) testing with a system realisation. Those stages are presented as large white blue-framed boxes in Fig. 4.11. The process involves four main iteration cycles (blue circles) taking information from green boxes as input and creating outcomes shown in blue boxes. The main artefacts of the process are the SGAM architecture model for ancillary services with ICT recommendations and ICT recommendations for system realisations.

The following subsections describe the details of each process stage.

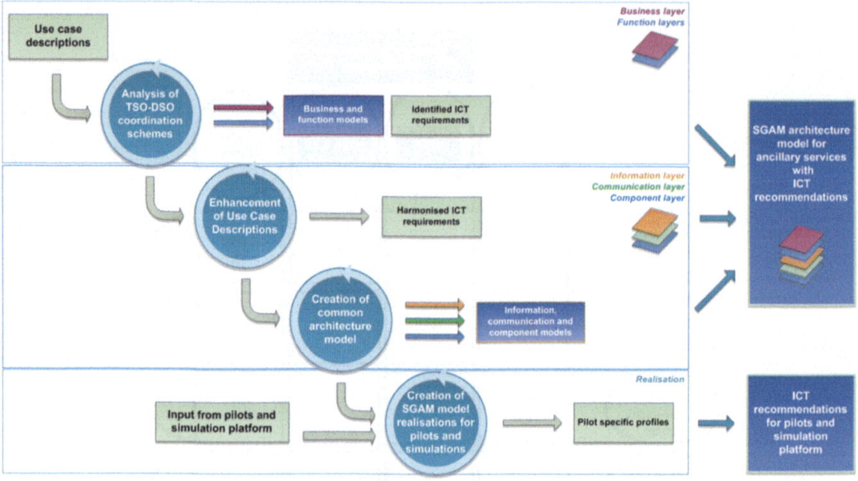

Fig. 4.11 Analysis procedure used for capturing ICT requirements and specifying the architecture design [19]

4.3.2.1 Stage 1: Classification of ICT Requirements

This stage captures and prioritises ICT requirements by analysing the five TSO-DSO coordination schemes and associated ancillary service use cases. This first stage focuses on business and function layers. The ELECTRA's use case template is used to identify business and system actors and their interactions. In the SmartNet project, the template was enhanced to include also functional ICT requirements related to [18]:

- Communication technologies
- Latency and security aspects
- Bandwidth
- Coverage
- Scalability
- Ownership
- Terminal density
- Interface flexibility
- OPEX and CAPEX costs
- Market characteristics
- Reliability
- Security (integrity, availability, confidentiality, authentication, non-repudiation)
- Data and communication protocols

At this stage, the identified requirements are still generic and business/function driven focusing mainly on market characteristics and interactions between business actors.

4.3.2.2 Stage 2: Harmonisation of ICT Requirements and Creation of the Common Architecture Model

The second stage refines the architecture design by extending the design and ICT requirements to information, network, and component layers. This involves complementing functional requirements with data structures and protocols. At this stage, suitable data and security-related protocols, e.g. EDI and IEC 62325, are investigated. The design covers physical system components and their interactions. Corresponding ICT requirements are mapped to the following requirement categories with own properties:

- Security
- Communication
- Latency
- Data protocol
- Device

Categorisation of ICT requirements is needed to ensure that the number of requirements remains manageable and that they can be aligned across all SGAM layers to form a common architecture model.

ICT systems evolve faster than energy systems making it difficult to choose optimal ICT solutions with long life expectancy for different parts of the energy system. As a result, system functions and their requirements tend to change more frequently in the future. To cope with this, the SmartNet project developed a parametrised SGAM realisation with Enterprise Architect (EA) architecture design tool. The parametrised model means that ICT requirements and their threshold values can be altered and their effects in different coordination schemes can be analysed.

The system actors, business actors, and interacting systems covering all TSO-DSO coordination schemes are presented in Table 4.7. As stated earlier, the proposed coordination schemes have a lot of similarities from the ICT's viewpoint. All the coordination schemes are assumed to utilise existing communications infrastructure and differ mainly from IT systems used for calculating market clearings and aggregations.

Figures 4.12 and 4.13 show information exchange between the system components in cases of common TSO-DSO and local AS market models. In the figure, arrows are showing directions of data flows. The thickness of the line indicates how many different types of messages are exchanged between system components. The actual number of sent messages (communication cost) depends on system realisation and information exchange interval. The data exchange can be periodical or event driven. The colour of the line shows whether the communication link is external (black) or internal (blue). The internal link is considered more reliable and secure, since it is managed by a single actor. Figure 4.12 shows that the most critical system component in the common TSO-DSO market model is market management system (MO MMS) operated by a common market operator and majority of the interactions are over external communication links.

Table 4.7 System actors in TSO-DSO coordination schemes

System actor	Business actor	System
MO MMS	Market operator	Market management system
TSO MMS	Transmission system operator	Market management system
TSO TS	Transmission system operator	Trading system
TSO EMS/SCADA	Transmission system operator	Energy management system/supervisory control and data acquisition
DSO MMS	Distribution system operator	Market management system
DSO local MMS	Distribution system operator	Local market management system
DSO TS	Distribution system operator	Trading system
DSO DMS/SCADA	Distribution system operator	Distribution management system/supervisory control and data acquisition
Aggregator TS	DER aggregator	Trading system
Aggregator EMS/SCADA	DER aggregator	Energy management system/supervisory control and data acquisition
CMP TS	Commercial market player	Trading system
CMP EMS/SCADA	Commercial market player	Energy management system/supervisory control and data acquisition
DER TS	Distributed energy resources	Trading system
DER EMS/SCADA	Distributed energy resources	Energy management system/supervisory control and data acquisition

In case of the local AS market model, the critical point is local DSO's market management system (DSO MMS). There exist more blue lines than in common TSO-DSO model, which indicates that more information can be transmitted over internal links. This improves the reliability and security of the overall system.

Requirements related to, e.g. latency, reliability, and security tend to change as energy systems, markets, and communication technologies evolve. Latency and security requirements are major factors when decisions to deploy wireless or wired connections are made. Data amounts exchanged in trading and resource control are not considered large, so the speed is not a critical factor. Wireless connections are more cost-effective and flexible, but wired connections offer more speed and reliability.

Figure 4.14 shows an example where the model parametrisation is used to assess the deployment possibilities of wireless communication technologies. The threshold values for latency and security in wireless connections are set as:

$$\text{Latency} \leq 100\text{ms} \ \text{ or } \ \text{Security level} \geq 4 \tag{4.1}$$

If none of the messages exchanged between two system components have stricter criteria, then a wireless option for the communication link can be utilised. The out-

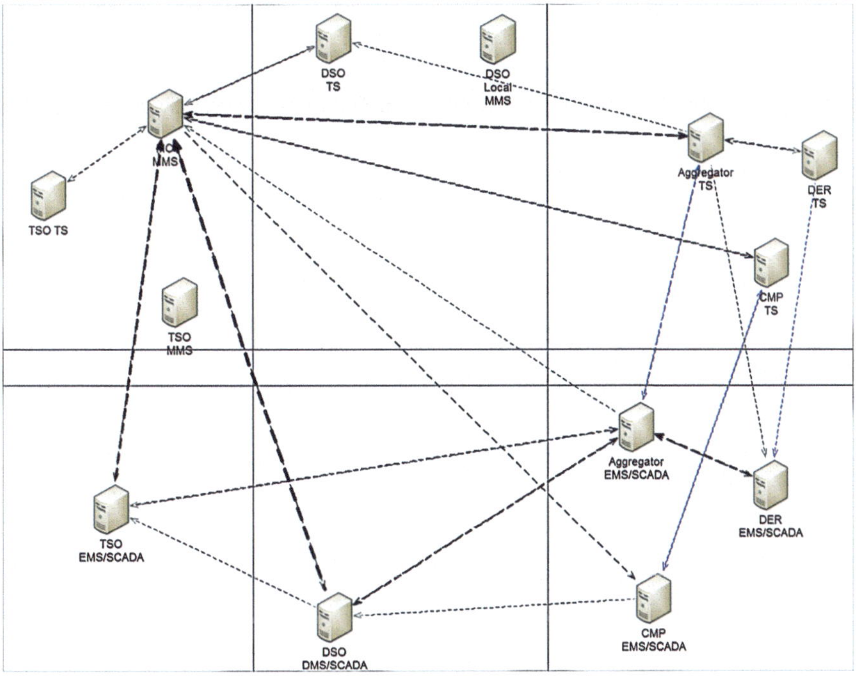

Fig. 4.12 Key interactions between system components in a common TSO-DSO market model

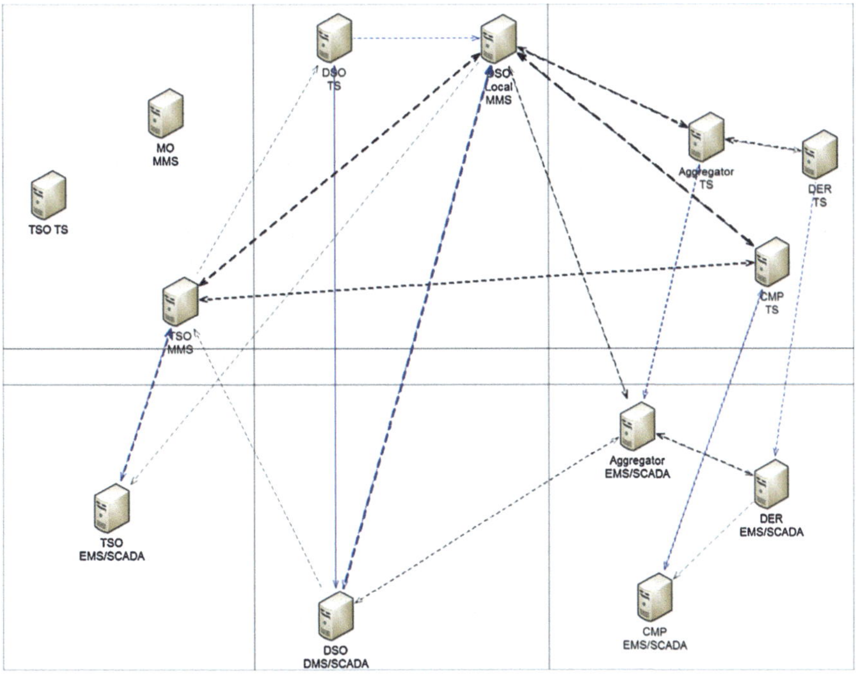

Fig. 4.13 Key interactions between energy system components in a local AS market model

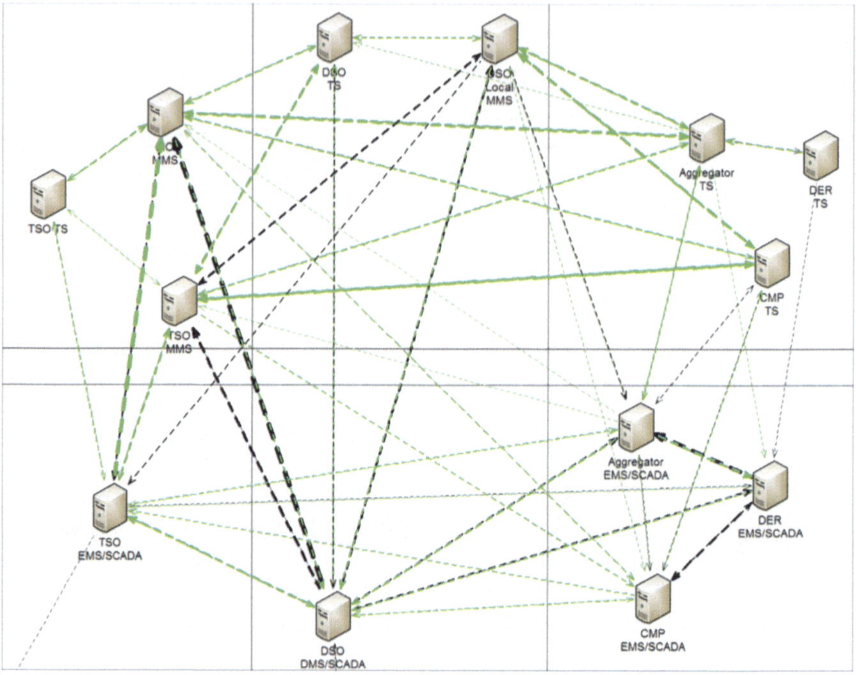

Fig. 4.14 Communication types: wireless connections are shown in green, and wired connections are shown in black

come is presented in Fig. 4.14, where wired connections are shown in black having more stringent requirements for latency and/or security and wireless connections in green with more relaxed requirements. The diagram indicates that wireless connections could be deployed more frequently. The wired connections are mainly needed for exchanging resource control information with SCADA systems in TSO, DSO, aggregator, and DER levels.

The advantage of using a parametrised model is that it enables to alter ICT requirements and threshold values and to compare their effects to different TSO-DSO coordination schemes. However, it is important to keep in mind that the outcome of the analysis depends on how precisely ICT requirements and threshold values can be defined for each connection in the target system. Therefore, ICT requirements need to be analysed systematically in all SGAM layers.

4.3.2.3 Stage 3: Testing with a System Realisation

The last stage of the analysis process is to utilise the system design in a real implementation. In the SmartNet project, one of the target systems was the Danish pilot described in detail in Chap. 6. This pilot was implemented to use summer houses with swimming pools for the provision of ancillary services. The electrical load used to heat water in a swimming pool is used for balancing the load. The

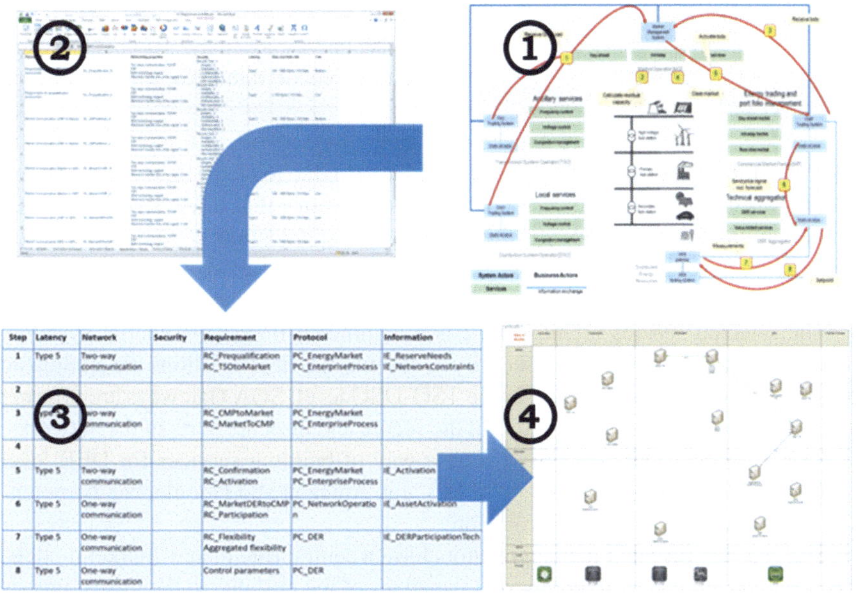

Fig. 4.15 A graphical representation of the analysis steps for the Danish pilot

coordination scheme of the pilot is combined "mixed shared balancing responsibility" and "common TSO-DSO" AS markets.

The following four analysis steps were taken (Fig. 4.15):

1. Defining the sequence of core information exchanges between pilot components utilising the conceptual reference model.
2. Selecting ICT requirements for each information exchange event from the ICT requirements catalogue generated during the overall system design.
3. Configuring the selected ICT requirements to match the pilot-specific requirements.
4. Creating a parametrised SGAM realisation to see which system components are needed and which information exchange links could be implemented over wireless technologies.

Although the chart in step 4 presents only few links (wireless connections in green and wired connections in black) between components, the created digital presentation includes different types of diagrams with ICT requirements covering all SGAM layers. This provided additional design support for the pilot system realisation is described in Chap. 6.

4.4 Conclusion

In this chapter, the importance of ICT in future energy systems is presented. Identification of ICT requirements is needed from business use cases to system components. From ICT's viewpoint, the presented five coordination schemes have a lot of similarities and existing data exchange protocols and standards, e.g. EDI and IEC 62325, form a good foundation for future energy systems. To enable cross-border interoperability, the common data and protocol standards should be used whenever possible.

Wireless technologies are expected to have a bigger role in future systems and could be utilised especially in the edges of the grid due to cost-efficiency and flexibility. This chapter presents also enabling technologies that offer new alternatives for communications. Choosing the optimal communication solutions, however, depends on several factors, e.g. regulation, business and market models, existing infrastructure, end-user requirements, and investment and operation costs.

The selected service architecture has a significant impact on energy systems' interoperability, flexibility, and security. Conventional ESB solutions are applicable in core parts of the systems, e.g. in the TSO-DSO level. SOA Gateways are offering additional security by enabling to split the secure and insecure parts of the networks, and microservices can be a cost-effective way of building services for DER level trading. In addition, global data providers, e.g. Amazon and Google, can be potential new stakeholders for data provision.

The analysis revealed that aggregators have a central role in new market models. The security threats are the highest at the edges of the grid due to the lowest investments to communication quality and security. Security aspects are adequately and systemically planned and implemented by DSOs, TSOs, and large aggregators. Regulatory support may be needed for small DERs' owners and aggregators that may not have sufficient competence or capital to invest on equipment and software to make their communication links and data secure.

The chapter presents a process for capturing ICT requirements and how parametrised SGAM model can support the system design and implementation. However, the outcome depends on how well the ICT requirements of the system can be identified in all SGAM layers and how much technological, economical, and regulatory uncertainties there exist.

References

1. 3GPP, Release 8, The Mobile Broadband Standard website (2014, September 24). Retrieved January 20, 2017, from http://www.3gpp.org/specifications/releases/72-release-8
2. 3GPP, Release 13, The Mobile Broadband Standard website (2015, September 17). Retrieved January 20, 2017, from http://www.3gpp.org/release-13
3. 5G PPP, 5G Empowering Vertical Industries. (2016, February 23). Retrieved January 20, 2019, from https://5g-ppp.eu/wp-content/uploads/2016/02/BROCHURE_5PPP_BAT2_PL.pdf
4. Analysys Mason, LPWA networks for IoT: worldwide trends and forecasts 2015–2025 (2016). Retrieved February 14, 2019, from http://www.gsma.com/newsroom/press-release/gsma-launches-low-power-wide-area-network-initiative-accelerate-growth-internet-of-things/
5. CEN-CENELEC-ETSI Smart Grid Coordination Group, Smart Grid Reference Architecture (2012, August 11). Retrieved January 20, 2019, from https://ec.europa.eu/energy/sites/ener/files/documents/xpert_group1_reference_architecture.pdf

6. N. Damsgaard, J. Helbrink, G. Papafthymiou, Study on the effective integration of Distributed Energy Resources for providing flexibility to the electricity system (2015)
7. K. Dinh, An Overview of Microservices Architecture (2015, May 1). Retrieved February 14, 2019, from http://khoadinh.github.io/2015/05/01/microservices-architecture-overview.html
8. ELECTRA, Deliverable R4.1, Description of the methodology for the detailed functional specification of the ELECTRA solutions (2015)
9. European Commission, Towards 5G (2019). Retrieved January 23, 2019, from https://ec.europa.eu/digital-single-market/en/towards-5g
10. P.M. Evjen, Wireless M-Bus module for AMR in the Netherlands. Metering International (2008)
11. Group, C.-C.-E.S. SGCG/M490/G_Smart Grid Set of Standards (2014)
12. ITU-R, Recommendation ITU-R M.2083-0, IMT Vision – Framework and overall objectives of the future development of IMT for 2020 and beyond. International Telecommunication Union (2015, September)
13. Nokia, Looking ahead to 5G, Workshop presentation (2014, August)
14. Nokia, LTE-M – Optimizing LTE for the Internet of Things, White Paper (2015). Retrieved January 20, 2019, from https://novotech.com/docs/default-source/default-document-library/lte-m-optimizing-lte-for-the-internet-of-things.pdf?sfvrsn=0
15. P. Panigrahi, LTE Vs HSPA+: Where is the future? (2012, August 9). Retrieved January 20, 2017, from http://www.3glteinfo.com/lte-vs-hspa-where-is-the-future/
16. U. Raza, P. Kulkarni, M. Sooriyabandara, Low power wide area networks: An overview. IEEE Commun. Surv. Tutor. 19 ss. 855–873 (2017)
17. SmartNet, D1.3: Basic schemes for TSO-DSO coordination and ancillary services provision (2016)
18. SmartNet, D3.1: ICT requirements specifications (2016)
19. SmartNet, D3.2: ICT architecture design specification (2017)
20. SmartNet, SmartNet website (2019). Retrieved January 7, 2019, from http://smartnet-project.eu/
21. TemaNord, *Nordic Data Hubs in Electricity System: Differences and Similarities* (Nordic Council of Ministers, Copenhagen, 2017)
22. UK Government Chief Scientific Adviser, *Distributed Ledger Technology: Beyond Block Chain* (Goverment Office for Science, London, 2016)

Chapter 5
Scenario Analysis

Marco Rossi, Julia Merino, Carlos Madina, Elena Turienzo, Harald Svendsen, Pirkko Kuusela, and Pekka Koponen

5.1 Introduction

The increase of renewable-based generation and the business potential for demand response is expected to gradually move the centre of mass of regulation reserves closer to the distribution system in the next future. This evolution will likely request the involvement of distribution system operators in the procurement and activation of flexibility, including the definition of new services dedicated to the cost-effective operation of the distribution network.

According to Chap. 2, these aspects can be managed by means of dedicated TSO-DSO coordination schemes. Each of them is characterized by certain peculiarities which make it the most promising one in given conditions. For this reason, the scenarios in which TSO-DSO coordination will be requested potentially include several key aspects to be investigated.

M. Rossi (✉)
Ricerca sul Sistema Energetico – RSE S.p.A., Milan, Italy
e-mail: marco.rossi@rse-web.it

J. Merino · C. Madina · E. Turienzo
TECNALIA, Derio, Spain

H. Svendsen
SINTEF Energy Research, Trondheim, Norway

P. Kuusela · P. Koponen
VTT Technical Research Centre of Finland Ltd,
VTT, Espoo, Finland

G. Migliavacca (ed.), *TSO-DSO Interactions and Ancillary Services in Electricity Transmission and Distribution Networks*,
https://doi.org/10.1007/978-3-030-29203-4_5

The construction of future scenarios is a complex task and requires numerous information sources, especially when a complete electrical system has to be developed in order to be simulated at capillary level. In fact, TSO-DSO interactions can be efficiently evaluated only by extending the conventional scenario analyses (normally limited to the detail of transmission network) to distribution systems too (for which much less information are normally available). This necessity also increases the difficulties in carrying out simulations aimed at identifying the main factors that influence the performance of TSO-DSO coordination schemes. In addition to the large model dimension (which includes both transmission and distribution grids), an enormous amount of power devices have to be simulated, taking into account the peculiarities of each of them in providing flexibility reserve.

The chapter summarizes the main procedure that can be adopted for the definition of likely future scenarios, particularly focusing on three reference countries: Denmark, Italy and Spain (Sect. 5.2). The main high-level assumptions are discussed and used in order to translate them in detailed datasets that can be implemented in a simulation environment (Sects. 5.3 and 5.4). In addition, precise indications to carry out simulations are reported, identifying the main strategies to be adopted for an effective simulation of the market, bidding and dispatching processes, together with the physical behaviour of the electricity network and connected flexible devices (Sect. 5.5). Finally, thanks to the simulation results for the hypothesized scenarios, a detailed cost-benefit analysis has been proposed, which identified the main scenario assumptions that influence the performance of the different TSO-DSO coordination schemes (Sect. 5.6).

5.2 Identification of Future Needs for Ancillary Services

The definition of scenarios to represent alternative images of how the European power system could develop in the future is a complex task that cannot be easily drafted. Several reports have dealt with the problematic, but the projections to the future clearly differ. They have been derived either by some international groups or they have been developed in specific projects. An analysis of different methodologies, considering strengths and weaknesses, can be found in [1].

As a result of this analysis, the key parameters of the e-Highway 2050 methodology can be selected to exhaustively describe future behaviour of the European power system. These include not only the degree of compliance with European Commission (EC) emission target and massive integration of renewable energy sources (RES) but also cross-border system expansion, demand response (DR) initiatives and storage devices that contribute to system balance and to reduce the need for grid capacity expansion and backup power.

5.2.1 *Expected Trends in Power Systems Affecting the Scenario Design*

5.2.1.1 Electricity Generation Mix

Up to 2030 the EU power generation mix is expected to change and the electricity produced by RES will likely increase from around 21% in 2010 to 44% in 2030 [2]. These forecasts of the EC are still conservative since some other sources even raise the number up to 89%. Wind energy, solar photovoltaic and biomass will increase their contribution (especially from onshore wind energy). Hydro and geothermal will remain constant, while, on the other side, nuclear energy will decrease from 27% in 2010 to 22% in 2030 even if some new power plants are expected and some others will be subject to life extensions. From all the RES, the intermittent generation (wind and solar) will be around 25%, emphasizing the need for flexibility in the system.

5.2.1.2 Energy Storage

Higher penetration of RES in distribution networks will increase the need of reserves to maintain system stability. Due to its characteristics, storage systems are suitable for ramping up/down to absorb small fluctuations caused by intermittency and forecasting errors.

The European Energy Storage Roadmap [3] shows a projection about the expected drop of storage prices. Additionally, it justifies the profitability of storage technology at utility level rather than centralized. In the 2030 horizon, storage will be fully integrated in power systems as resources of flexibility for several ancillary services.

5.2.1.3 Demand Response

Participation of customers is being promoted by current development and challenges such as the deployment of advanced metering infrastructures. DR is also currently provided by the big industrial loads mainly via interruptible load programs connected with the transmission system operator (TSO) or with a balancing responsible party (BRP). Even though there is a big market potential for DR participation, its integration is still small due to existing regulations in most wholesale markets. For example, the minimum volume and response duration required by existing markets make it difficult for small users to participate in the day-ahead, intraday or ancillary services markets. Additionally, there are still many challenges to be solved for the effective integration of DR. They can be summarized into four main points:

- Clear definition of the aggregator role and responsibilities
- Impact on the BRPs' portfolio due to activation by aggregators
- Safe and reliable data flows between agents
- Clear and smart control and market mechanisms

Regardless all the barriers that are being removed, a potential demand response of 160 GW has been predicted for Europe by 2030 [4].

5.2.1.4 Enhancement of Cross-Border Interconnections

The sharing and exchanging of balancing reserves are expected to bring potential benefits for the power system, increasing the competition in the markets, providing cheaper access to the resources and reducing costs for end users. Regardless of the model implemented, one of the main challenges for cross-border exchange is how to allocate the available transmission capacity at interconnectors, in order to make it in the most economically feasible way, and which allows an optimal utilization of the available transmission capacity between the national transmission systems. Since cross-zonal capacities are a limited resource, capacity reservations are quite controversial, because they mean that less cross-zonal capacity will be available for day-ahead and intraday trading. Because it is not possible to predict in advance how much of the contracted reserves will actually be used at a certain future time, there is a risk that the valuable transmission capacity will not be used. Since the interconnectors are likely to be congested in hours with high price differentials between bidding areas, the value of the trade lost is expected to be high.

5.2.2 Design of High-Level Scenarios

A scenario can be described as a possible development or projection of different issues into the future. It describes the relationship between several variables in such a way that modifying a parameter brings changes to all related aspects. The scenarios have to offer insights into the future and challenge the conventional wisdom about it, showing the way to be followed to get there from the present situation.

As mentioned, scenarios can be defined on the basis of the analysis carried out within the e-Highway 2050 project [5]. A major reference for the application of this methodology consists of [6]. The application can be summarized in the four steps presented in Fig. 5.1.

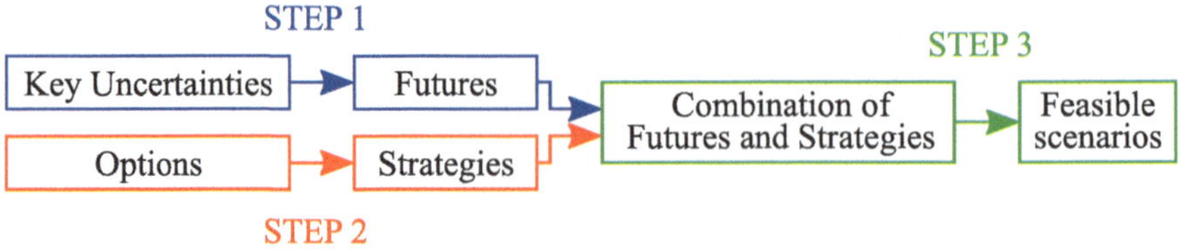

Fig. 5.1 Four-step methodology for high-level scenario design

5.2.2.1 STEP 1: Definition of Key Uncertainties and Futures

Uncertainties are those parameters that are beyond our control and thus, they pose serious doubts about their feasible evolution (such as the fuel prices, the economic growth, etc.). Considering ENTSO-E European Generation Adequacy Outlook [7], the following key uncertainties have been selected:

- Economic and financial conditions
- European framework
- Technological breakthroughs
- Energy efficiency breakthroughs
- Usage of demand response potential
- Vehicle to grid initiatives
- Generation mix
- Transmission and distribution grids
- Smart grid technology implementation

In the present case, the four ENTSO-E visions shown in Fig. 5.2 have been considered as possible *futures*, highlighting the main characteristics, including the ones that are shared among different visions.

Vision 1, being the most pessimistic, is the only one that does not consider the electric vehicle for V2G initiatives or DR. Visions 1 and 3 share a weak European framework so they consider the generation mix based on national policies, with no changes in their grid connections. Instead, visions 2 and 4 rely on the improvement of grid connections based on a strong European framework. Visions 1 and 2 present less financial and economic conditions and thus, they are delayed in the achievement of EC 2050 targets [8], while visions 3 and 4 are expected to reach them.

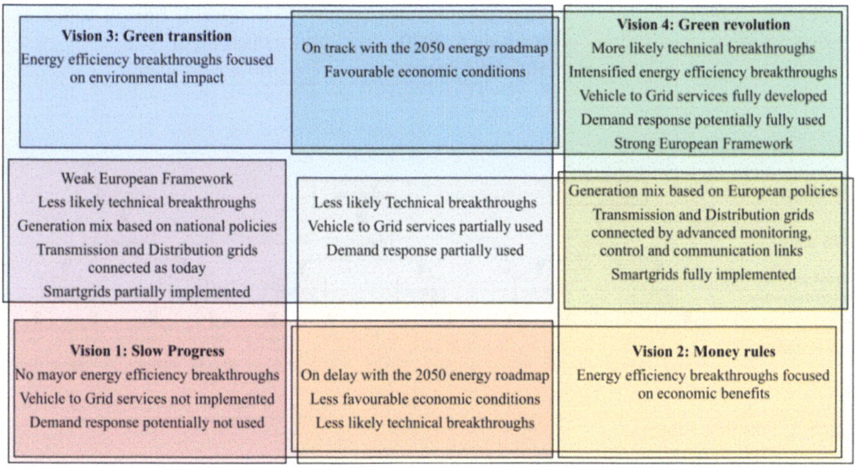

Fig. 5.2 ENTSO-E visions

5.2.2.2 STEP 2: Definition of Options and Strategies

In the second step, the *options* are defined and combined to define possible *strategies*. Unlike uncertainties, *options* are selected as a result of a decision-making process after considering some assumptions. For the definition of the strategies, the scenarios proposed by [9] can be considered as reference, which main characteristics are listed in Table 5.1.

As mentioned above, together with the EC emission targets and RES share, two additional *options* have been selected for the design of the strategies, as both aspects are essential to provide the system with increased flexibility, both at grid level and on the demand flexibility side: the cross-border interconnection status and the DR/storage presence. On the one hand, the interconnection to neighbouring systems has been evaluated as a driver for RES integration, reducing the need for backup power. On the other hand, DR helps in the reduction of grid capacity expansion and backup power. The criterion for RES share was the degree of compliance with the requirements to achieve the EC emission targets by 2050 and each of the levels defined corresponds to one of the scenarios in Table 5.1 (lower = business as usual, required = on track, higher = high renewable). Figure 5.3 shows all the possible *strategies* obtained by combining the considered *key options*.

Table 5.1 Summary of scenarios proposed by the EC within [9]

Business as usual	On track	High renewables
Non-fulfilment of current ENTSO-E plans, National Renewable Energy Action Plans up to 2020 and EC goals RES targets not fulfilled (26% share)	Compliance with the EC targets of reducing emissions by 2030 Fulfilment of current ENTSO-E plans and National Renewable Energy Action Plans up to 2020 50% RES share	Compliance with the EC targets of reducing emissions by 2030 Fulfilment of current ENTSO-E plans and National Renewable Energy Action Plans up to 2020 60% RES share

Fig. 5.3 Possible strategies

5.2.2.3 STEP 3: Combination of Possible Scenarios and Selection of the Feasible Ones

The next step consists of the linking of the considered *futures* and *strategies* to draft a set of possible and coherent scenarios. Table 5.2 shows the possible combination of scenarios, highlighting the incompatibilities that are occurring with the resulting combinations. For example, *strategies* that assume the achievement of emission targets in 2030 (strategies 5–12) are not compatible with the ENTSO-E visions 1 and 2. The ones based on the non-compliance with that targets are not compatible with vision 3 and vision 4. In total, after considering all the incompatibilities, six feasible scenarios (SCN) have been obtained which can be practically merged in four:

- Scenario 1 (SCN1) results from the combination of low shares of RES with ENTSO-E vision 2.
- Scenario 2 (SCN2) results from the combination of low shares of RES with ENTSO-E vision 1.
- Scenario 3 (SCN3 and SCN5) results from the combination of RES shares equal (or higher) than the EC targets with ENTSO-E vision 4.
- Scenario 4 (SCN4 and SCN6) results from the combination of RES shares equal (or higher) than the EC targets with ENTSO-E vision 3.

5.2.3 Mapping of European Countries to the High-Level Scenarios

In order to evaluate the performance of the TSO-DSO coordination schemes proposed by Chap. 2 on the different 2030 scenarios, three European countries have been investigated by the European H2020 project SmartNet: Denmark, Italy and

Table 5.2 Mapping of the scenarios on the basis of the considered strategies and futures

		Strategies											
		EC emission target not achieved				EC emission target achieved				EC emission target achieved			
		Lower RES share				Required RES shared				Higher RES shared			
		1	2	3	4	5	6	7	8	9	10	11	12
Futures	Vision 1	CBI DR	CBI	DR	**SCN 2**	RES CBI DR	RES	RES DR	RES	RES CBI DR	RES CBI	RES DR	RES
	Vision 2	**SCN 1**	DR	CBI	CBI DR	RES	RES DR	RES CBI	RES CBI DR	RES	RES DR	RES CBI	RES CBI DR
	Vision 3	RES CBI	RES CBI DR	RES	RES DR	CBI	**SCN 4**	DR	CBI	CBI DR	CBI DR	**SCN 6**	DR
	Vision 4	RES	RES DR	RES CBI	RES CBI DR	**SCN 3**	DR	CBI	CBI DR	**SCN 5**	DR	CBI	CBI DR

CBI incompatible for cross-border interconnection, *DR* incompatible for demand response and storage, *RES* incompatible for renewable energy share, *SCN* feasible scenario

Table 5.3 Flexibility resource total availability hypothesized by the 2030 scenarios for Denmark, Italy and Spain

		Year 2030		
		Distribution (MW)	Transmission (MW)	Total (MW)
Wind T.	Denmark (DK1)	853	2,438	3,291
	Italy	1,261	3,041	4,303
	Spain	5,317	2,907	8,224
PV	Denmark (DK1)	267	0	267
	Italy	6,945	89	7,034
	Spain	5,451	70	5,521
Stationary storage: battery	Denmark (DK1)	0	0	0
	Italy	19	123	142
	Spain	4	0	4
Stationary storage: hydro	Denmark (DK1)	0	0	0
	Italy	817	6,470	7,287
	Spain	11	6,968	6,979
Stationary storage: flywheel	Denmark (DK1)	0	0	0
	Italy	0	0	0
	Spain	1	0	1
Mobile storage	Denmark (DK1)	6,000	0	6,000
	Italy	285	0	285
	Spain	2	0	2
CHP	Denmark (DK1)	990	825	1,815
	Italy	4,841	13,020	17,861
	Spain	3,719	3,100	6,819
TCL	Denmark (DK1)	306	72	378
	Italy	652	0	652
	Spain	772	0	772
Load shifting	Denmark (DK1)	228	0	228
	Italy	49	0	49
	Spain	37	0	37
Load curtailment	Denmark (DK1)	0	0	0
	Italy	394	0	394
	Spain	0	0	0
Industrial processes	Denmark (DK1)	119	0	119
	Italy	548	137	685
	Spain	72	287	358

Spain. For each of them and on the basis of the selected scenarios, the project also hypothesized the amount of resources capable of providing flexibility services (Table 5.3).

Denmark is expected to host large amounts of RES, higher than the required one for the fulfilment of the 2030 emission targets. The general goal consists of achieving, by 2030, a 40% cutback of the greenhouse gas emissions existing in 1990 and a CO_2 reduction in the power sector of 54 ÷ 68% [10]. The Danish Government is going even one step further, planning to reach the 40% reduction of greenhouse gases by 2020 [11]. This means that the RES share installed in Denmark is bigger than the required fulfilment of emission reductions, meaning that it is aligned with European policy schemes. This way, it can be clearly stated that scenarios 1 and 2, with a lack of RES, will be never representative of the Danish case.

The demand response implementation, instead, is foreseen to be still low in 2030 since current reserve mixes are expected to be available and at competitive prices. However, since most of the markets will allow the participation of DR, flexibility will be likely included in ancillary services reserves [12].

Concerning the cross-border interconnection status, in October 2014, the European Council [13] called to all European countries to reach the 15% interconnection target by 2030. The analysis done by ENTSO-E [14] shows the fulfilment of the interconnection targets by Denmark, no matter the vision. The match of the achievement of the RES goals together with the interconnections status clearly justifies the belonging of Denmark to the scenario SCN3.

In *Italy*, the perspectives are still unclear. The more updated and valuable forecasts were published in a joint report written by ENEA and RSE for the Italian Ministry Economic Development [15]. Having considered the recent (and still growing) increase of RES, vision 3 of ENTSO-E seems to be the most adherent to the Italian situation, with a RES penetration slightly below 50%. According to this assumption, the installed RES capacity is expected to reach around the 46.5% (70 GW) of the total energy mix, which will maintain only a 4% (7 GW) of hard coal-fired power plants and the 25% (37 GW) of combined cycle gas turbines. Italy, as Denmark, will reach the interconnection goals. For all the reasons abovementioned, the hypothesized scenario SCN4 is considered the one that better suits in the Italian situation.

Concerning *Spain*, the situation is more pessimistic. In the 2030 framework there are not envisioned further investments in RES. This does not necessarily mean the target of CO_2 reductions is not expected to be fulfilled but rather that emission target mechanisms will be likely adopted [16]. Due to the historical situation between Spain and France (boundary subjected to significant congestions), the compliance with the interconnection target of 10% is likely unfeasible, as Spain is nowadays still far away from the target. In March 2015, French, Spanish and Portuguese governments signed the Madrid Declaration with the EC. This declaration showed the will to develop four joint projects to increase the capacity between France and Spain to 8 GW. Nonetheless, it is still not considered enough, even though this investment effort will improve very much the interconnection ratio of Spain [14]. Due to the remote possibility of Spain to meet the interconnection goals, the most probable scenario for Spain is SCN2.

5.3 Mapping of National Scenario Down to the Electricity Network Nodes

The high-level scenarios described above provide information at national/regional level which, in order to be investigated in terms of provision of ancillary services, further indications are needed. In particular, the physical characteristics of individual devices and their position on the electricity network are fundamental inputs for the construction of detailed scenarios to be simulated.

5.3.1 Assumptions for the Definition of the Detailed Scenario

Identifying the geographical position of the devices is an intermediate step for the construction of the electrical model. In fact, numerous sources can be used in order to make deductions on possible allocation of power devices over the national territory and, once this information is available, they can be easily mapped on the electricity grid.

In some cases, source data already provide the connection point on the transmission network. When this information is available, devices can be already mapped precisely on the electrical grid (this particularly happens for large generation and load units). In other cases, grid connection points can be generally determined on the basis of the shortest distance between transmission network node and hypothesized geographical position.

5.3.1.1 Conventional Generators

Conventional generators consist of generation units having the power output controllable and a storable energy source (fuel). This includes conventional fossil fuel power plants, biofuel plants and hydroelectric plants with water reservoirs (run-of-the-river and pumped hydropower are classified differently since the first one has not storable fuel while the second one is treated as storage unit).

Several sources can be used in order to generate the dataset of conventional generators and most of them can be freely consulted [17]. More detailed databases, including also exhaustive information on power plant characteristics and dismissal expectation (particularly important from a scenario design perspective), are also available on private repositories [18]. Table 5.4 reports the power capacity of each considered generation technology assumed for the considered scenarios.

Concerning the production profiles, historical data can be easily found [17]. However, having considered that demand and generation mix is expected to be significantly different in 2030, their power output has to be recalculated in order to match the actual demand.

5.3.1.2 Non-programmable Renewable Generation

Wind turbines, solar and run-of-the-river power plants differ from conventional generators in the scenario definition since their energy source is not storable. The availability of power is determined on the basis of the actual weather conditions and the non-converted wind/solar radiation/water is lost energy.

Table 5.4 Expected installed generation capacity for the considered countries on the basis of the 2030 scenario assumptions

	Power capacity (MW)		
	Italy (Vision 3)	Denmark (Vision 4)	Spain (Vision 1)
Biofuels	0	1,460	0
Gas	37,993	3,746	24,948
Hard coal	7,056	410	5,900
Hydro	23,535	9	23,450
Lignite	0	0	0
Nuclear	0	0	7,120
Oil	1,386	735	0
Others non-RES	10,160	0	10,480
Others RES	10,750	260	2,400
Solar	40,400	1,405	16,800
Wind	18,990	12,825	35,750

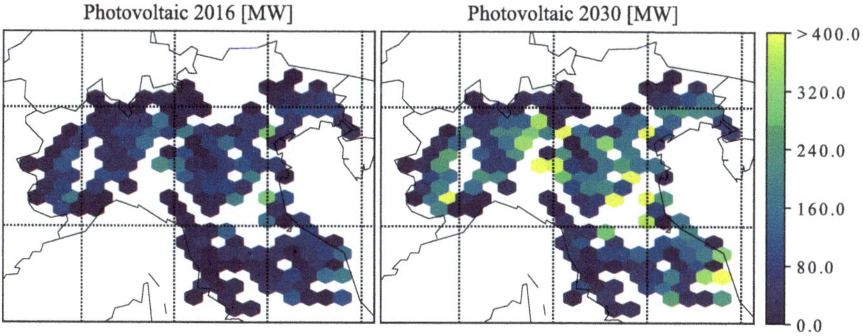

Fig. 5.4 Comparison between the current (2016) geographical distribution of PV generation in Italy and the one assumed in the 2030 scenario

Installation of renewable power plants is monitored in several countries and the information on geographical position, rated power and technology can be typically extracted for each power plant (or small-scale aggregation of them) from dedicated databases [19–21].

The mentioned data sources refer to the current situation which has to be scaled up in order to match the high-level scenario hypotheses. Within the considered countries, the power plants of a given technology have been assumed to be already located in areas where energy sources can be largely and efficiently exploited. For this reason, the capacity of each technology has been matched with the scenario predictions by proportionally increasing the number of devices within the areas where they exist already (Fig. 5.4 reports the illustrative case of PV expansion in Northern Italy). Having considered that no major technology updates are expected in the short term, the additional power plants have been assumed to have same physical characteristics (power capacity distribution) of the current ones.

Another fundamental input for the simulation of these devices consists of the power profile. As anticipated above, the power output of each technology is depending on different (but potentially correlated) weather variables. In order to use realistic profiles, taking also into account possible correlations, the power output of existing units can be extracted [17, 20, 22] or deduced from weather station measurements over the same time interval.

5.3.1.3 Combined Heat and Power Generation

The heat demand and power profile for combined heat and power generators have been assumed to be identical to the current one since the high-level scenarios do not foresee any capacity increase of this technology. The only exception is represented by Denmark for which, in the hypothesized 2030 situation, small power plants (located at distribution level) will be completely replaced by heat pumps.

Typically, information on existing units can be deduced from the same data sources adopted for conventional generators [18] and often they can be directly located on the transmission network node they belong.

Even in this case, the power profile can be somehow related to the weather conditions of a particular day, resulting in a correlation with the power outputs adopted for renewable generation. For this reason, the use of real profiles (e.g. downloaded from [17]), extracted for the same time ranges adopted for renewable energy resources, is recommended.

5.3.1.4 Storage-Based Devices

According to the discussed high-level scenarios, the technologies capable of performing storage functions mostly consist of pumped hydro power plants and electric vehicles. Some stationary battery storage units are expected too, particularly in Denmark.

Numerous information for pumped hydro and electrochemical battery power plants can be extracted from [23] and since the current situation is not expected to significantly evolve up to 2030, the same location of current power plants is adopted.

Concerning electric vehicles, the impact on regulation reserves is expected to be significant and different assumptions have to be made in order to consider the most relevant variables (power/energy capacity, driving patterns, etc.). Depending on the type of use of a specific electric vehicle, hypotheses on the availability of flexibility can be made. In particular, three typologies of driving pattern (Fig. 5.5) are expected to be the most representative samples [24]:

- Family cars, mostly available during the night, with some charging peaks during the day
- Commuter cars, constantly connected to the grid, except during the early morning and late afternoon when commuters are travelling
- Taxis, available during the night

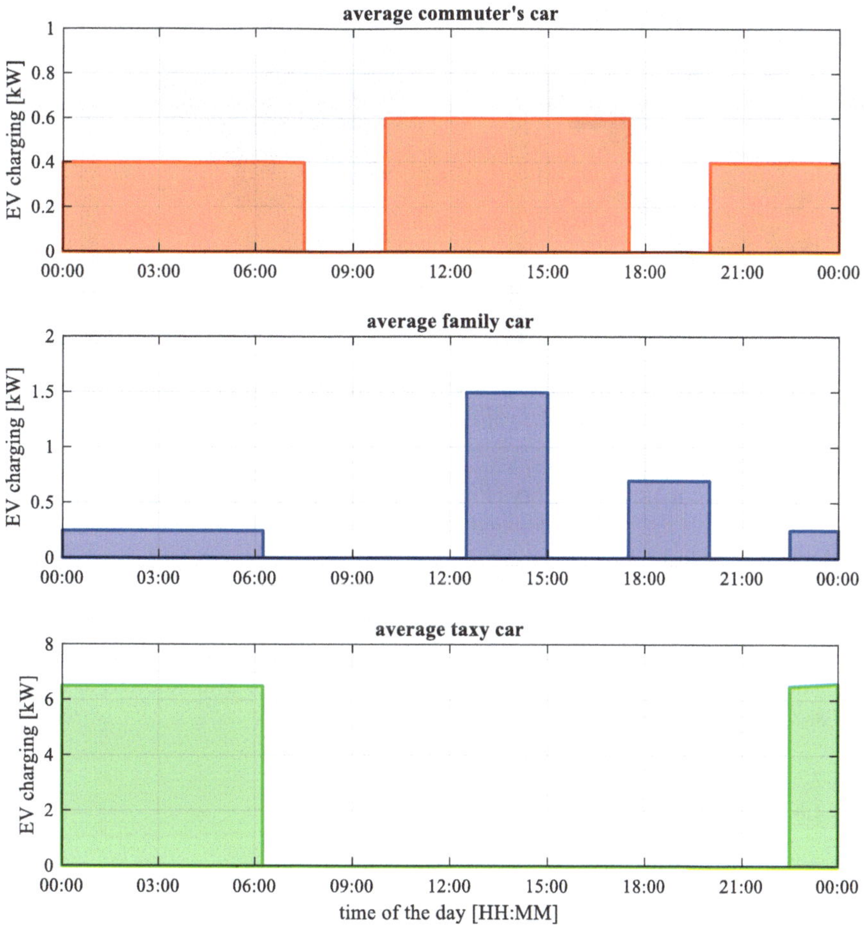

Fig. 5.5 Charging profiles of electric vehicles for family, commuters' and taxi cars

Different assumptions can be made in order to hypothesize the geographical location of electric vehicles during the charging phase. One simple possibility consists of using population statistics [25], considering both population volumes and densities for the deduction of the amount of vehicles connected to the same network node (which is identified as the closest bus to the identified geographical area).

5.3.1.5 Thermostatically Controlled Loads

According to the hypothesized high-level scenarios, the flexibility provided by thermostatically controlled loads is expected to be relevant and mostly provided by domestic dwelling and district heating systems. This means that, also for these devices, population statistics [25] can be used in order to deduce their future

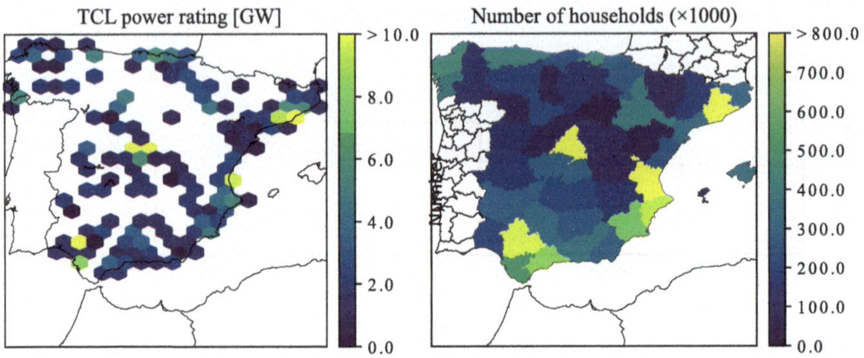

Fig. 5.6 Expected geographical distribution of thermostatically controlled loads compared to the predicted 2030 number of households in Spain

position over the national territory (Fig. 5.6). In addition to this source of information (since not all the heating demand will be satisfied by heat pumps) the amount of devices is deduced by comparing the national total power capacity (foreseen by the high-level scenario) with the nominal power and energy demand of heat pumps (which can be calculated by using simple thermal models of buildings and apartments – see Chap. 3). Then the amount of controllable devices (the ones offering their flexibility to the market) is taken as the percentage of available DR assumed for each country (reported in Table 5.3).

5.3.1.6 Atomic Loads

Atomic loads consist of devices with no flexibility on the consumption profile, except for the possibility of being deferred in time. Typically, domestic appliances (dishwashers, washing machines, tumble dryers) have good potential in terms of ancillary services provision once aggregated and optimized in order to provide the requested power flexibility.

The number of devices can be determined by looking at the expected amount of households over the considered territories which, again, can be deduced from population statistics [25]. In addition to households, ownership levels can be taken into account and a possible source of data is provided by the REMODECE project [26]. Finally, in order to build the scenario, assumptions on the number of actual flexible devices have to be taken on the basis of the amount of DR expected for each country.

At this point, realistic consumption profiles have to be taken and assigned to each simulated device. For instance, Virginia Tech has published power profiles measured on actual domestic appliances [27]. These profiles, once combined with the time instants in which the devices are likely to be activated [28], represent a realistic baseline situation for the considered scenarios (Fig. 5.7).

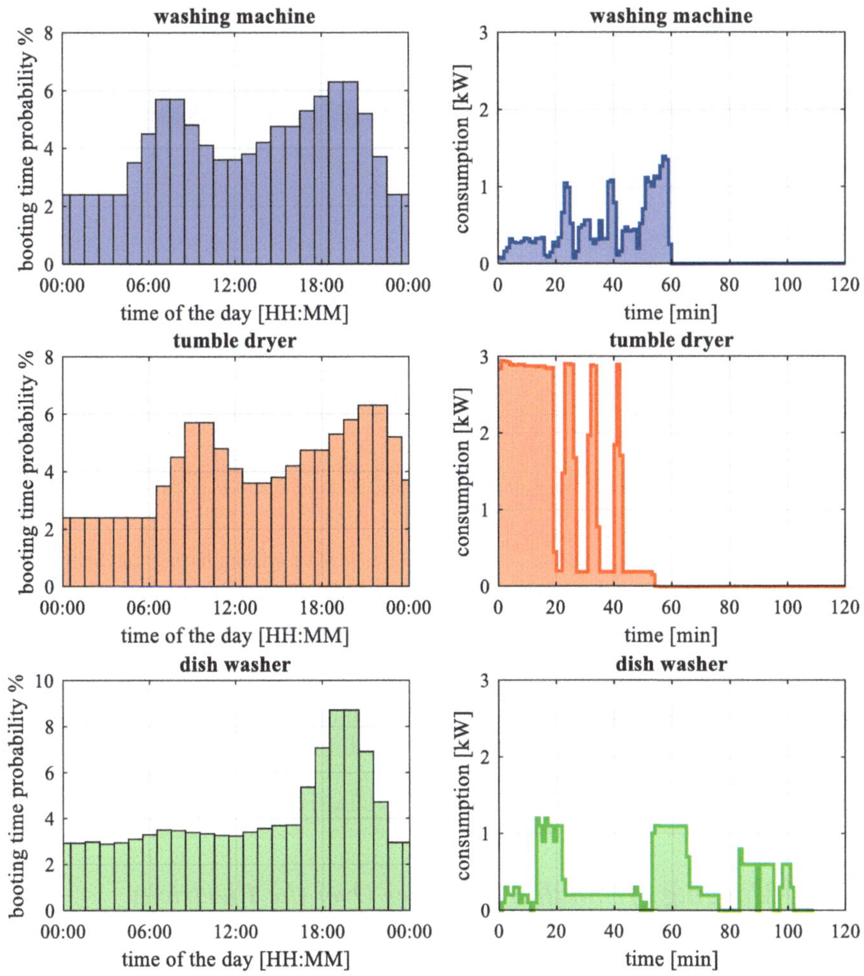

Fig. 5.7 Power profiles of actual domestic appliances and probability distribution of booting

5.3.1.7 Remaining Non-flexible Devices

Depending on the considered country, different evolutions of the demand profiles can be expected by 2030. Projections for this time horizon can be found for the correspondent ENTSO-E vision in [7]. Of course, once extracted, the portion of flexible loads has to be removed from the total demand curve (Fig. 5.8), thus obtaining the portion of non-flexible consumption (which can be bidirectional in case non-flexible generation is included). Again, the power profile can be distributed over the considered territory according to population or industrial evolution statistics [25].

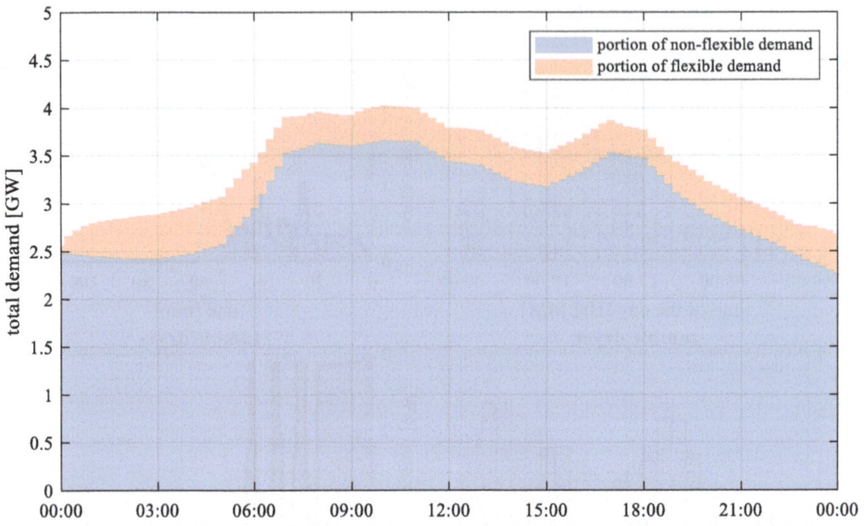

Fig. 5.8 Expected demand in Denmark according to ENTSO-E vision, having highlighted the portion of flexible and non-flexible demand (Denmark 2030)

Since the portion of non-flexible demand is subjected to uncertainty, forecasting errors have to be modelled for a correct simulation of the selected scenarios. The ENTSO-E transparency platform [17] reports both the day-ahead and actual profile of demand for each European country and this information can be used in order to reproduce realistic forecasting errors.

5.3.2 *Mapping of Simulated Devices Over the Electricity Network*

Once the set of devices to be simulated has been defined, including their possible geographical position, assumptions on their connection to the electricity grids have to be made (except for the cases in which this information is available already).

The first step to be carried out consists of mapping all the devices over the transmission network (even if they are supposed to belong to the distribution grid – Fig. 5.9). Transmission grids, in fact, represent the perfect link between the electricity model of the system and the geographical information available for each device, while the available distribution models are less likely including geo-topological data. The following assumptions can be considered a reasonable simplification for the construction of the electric scenario:

- The geographical information of transmission network nodes can be likely found on public repositories/websites [29–31]. Sometime these repositories already include the electrical model too [30]; otherwise they can be deduced from the physical characteristics of lines and cables when available [31].

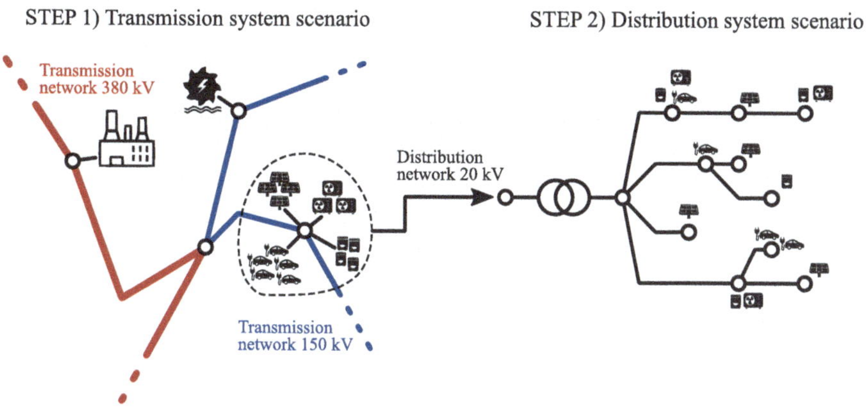

Fig. 5.9 Procedure for the mapping of energy resources on transmission (step 1) and distribution (step 2) network

- The devices having already information on the correspondent transmission network node are assigned to it.
- The remaining devices are connected to the closest free node. Transmission and distribution loads/generators can be easily distinguished by looking at their nominal power (10 MW typically represents a reasonable threshold) and possibly assigned to different nodes.

Thanks to this approach it is already possible to have a realistic electrical model of the transmission system. Simulations can be carried out already in order to see if the adopted network satisfies the system requirements (adequacy of line transfer capacities, voltage levels, etc.) and apply adaptation if needed. If the scenario is not too remote in future, 10-year reports on transmission network planning [32–34] can be used.

5.3.3 *Distribution Network Scenario*

At this point, the process described above returns a transmission system characterized by two node typologies. The first one consists of buses connected to few transmission power units, while the second one includes nodes connected to several distribution devices that have to be mapped on the correspondent distribution grid (Fig. 5.9). As anticipated above, information on the distribution networks typically are not publicly available and several assumptions have to be taken in order to define a realistic scenario for lower voltage levels.

In particular, the grid topology plays an important role for the definition of the distribution scenario. The main characteristics depend on the territory conformation, the amount and typologies of devices, and they have an impact on the kind of congestions (overloading/under-overvoltage) that might occur. For some countries, typical structures of distribution networks are publicly available for different typology of loads/territory (urban/rural/industrial) [35]. Otherwise automatic procedures to generate realistic distribution network topologies can be applied [36]. Most of them are limited to medium-voltage infrastructure which, for the level of details available, neglecting low-voltage infrastructure can be considered a reasonable compromise between accuracy and model complexity.

Since the position of the devices, which are hypothesized to be connected to a specific distribution network, cannot be known a priori, the SmartNet project defined a procedure based on the random allocation of power resources on the distribution grid nodes [24]. Possible positions of units are iterated by means of a Monte Carlo simulation and the ones fulfilling the following requirements are selected as feasible distribution network scenario (Fig. 5.10):

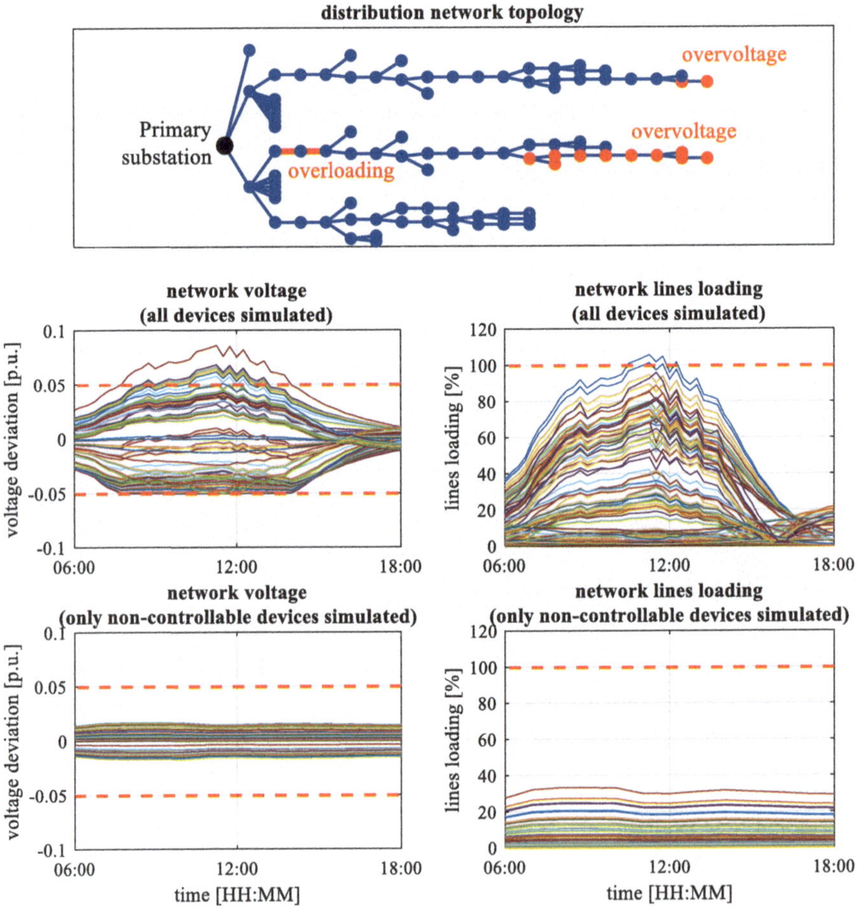

Fig. 5.10 Profiles of the electrical quantities for an illustrative distribution grid. The network is simulated with and without controllable resources. Under-voltage events are assumed to be managed by means of tap-changing transformers

- All the non-controllable resources can be safely supplied with or without the intervention of local flexibility.
- The operation and intervention of flexible resources can cause network congestions (this methodology guarantees that any critical situation can be solved by the market/DSO with the use of flexibility).
- Congestions are occurring for a limited amount of time (few hours per day) and severity (e.g. maximum +50% of loading/voltage capacity).

5.4 Detailed Scenario Dataset

At this point of the process, all the devices result to be connected to the related transmission/distribution network, and the electrical models of each scenario are completed. As described within the previous section, most of the simulated power units have a fixed baseline power generation (renewable-based generators) and consumption. These profiles are assumed to be the results of energy markets (day-ahead and intraday sessions), during which also the scheduling of conventional generators is decided.

5.4.1 Selection of the Reference Simulation Dataset

The complexity of the detailed scenarios, which include both transmission and distribution network together with hundreds of thousands of devices, makes the simulation of long periods (e.g. 1 year) impracticable. For this reason, for each of the simulation scenarios, few reference days have been selected in order to represent the most recurrent operating conditions. According to [24], the selection of these days can be based on the related availability of wind and solar generation (Fig. 5.11),

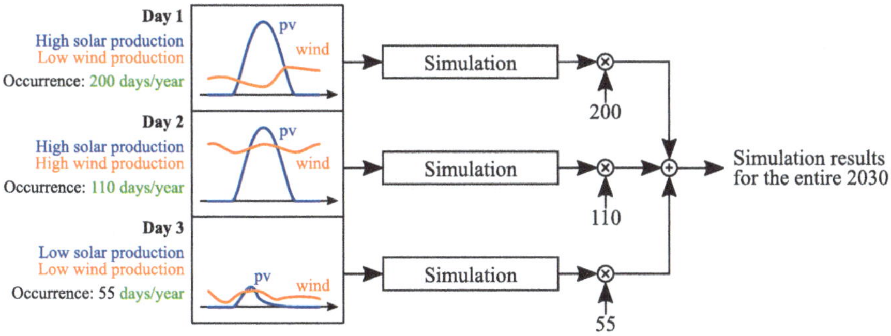

Fig. 5.11 Simulation strategy for the achievement of a simulation result set related to the full year of the hypothesized 2030 scenario

which are the resources subjected to the highest variability during the year. Other quantities (e.g. the power profile of conventional generators and loads baseline) are fixed or depend on the residual load/generation balancing.

5.4.2 Previous Market Data (Conventional Generators Baseline)

Since the hypothesized scenarios refer to a future situation, historical results of energy markets would not likely match it. For this reason, a dedicated simulation of the energy trading markets has to be carried out in order to obtain the missing profiles and the price of the energy according to the activated resources and interactions with neighbouring countries. This can be done by using PowerGAMA tool [37], a linearized optimal power flow, capable of simulating day-ahead market sessions having considered:

- Realistic day-ahead forecast of renewable generation (wind, solar, small-scale hydro)
- Day-ahead forecast of (non-controllable) demand
- Flexible (storage base, atomic, thermostatically controlled, curtailable) load profiles equal to their baseline
- Low cost for CHP units to ensure that thermal demand is satisfied
- Power losses added to the model as additional load

The output of the tool consists of the power profile of each conventional generator and the nodal price (defined for each transmission bus) for all the considered time steps. In order to obtain a single energy price for the energy, nodal prices are averaged (weighting them with the demand of each node). Marginal costs for generators depend on their technology and can be classified in non-fuel costs, fuel costs and CO_2 penalties. Realistic figures of them are provided by the project OffshoreGrid [38].

Figure 5.12 reports the energy market results of the considered Spanish case. Assumption on the energy price of neighbouring countries has been made in order to model their impact on the energy trading.

In theory, the final scheduling is decided by intraday markets which would add significant complexity to the scenario dataset. According to Chap. 3 simplifications can be adopted by introducing intraday trading discomfort costs to the flexibility bids of each device. If included, the adjustment of energy prices can be computed offline on the basis of the prices returned by the PowerGAMA tools and as a function of the imbalance occurring in correspondence of the intraday sessions (mostly due to forecasting error of renewables and demand).

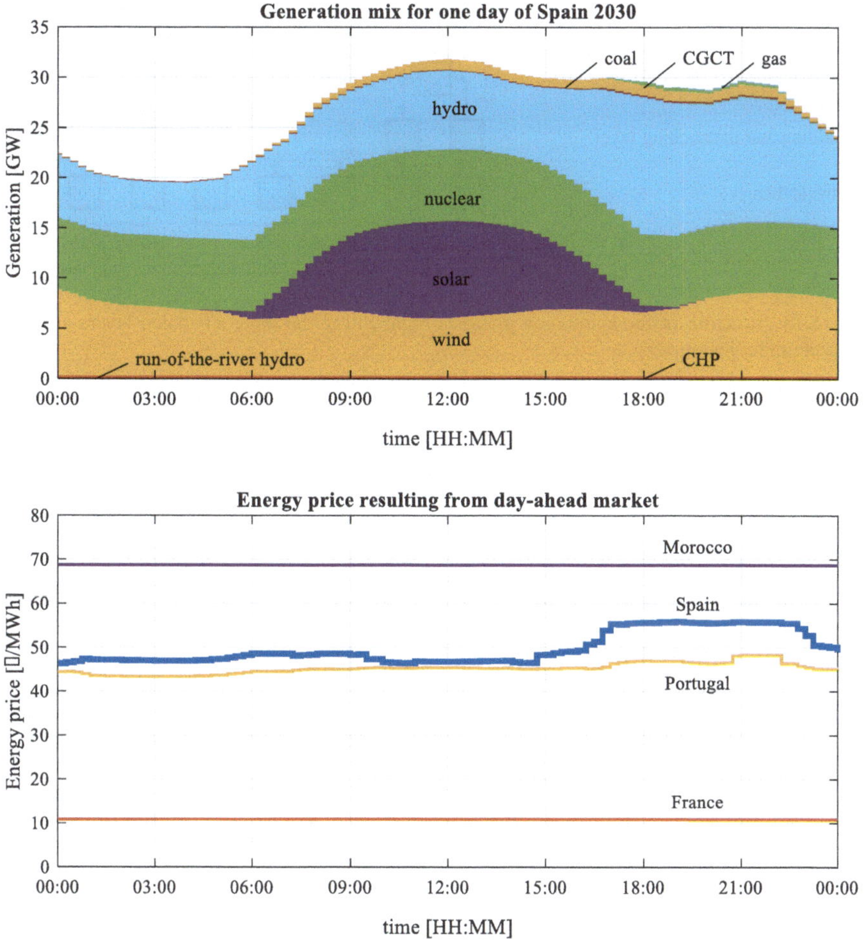

Fig. 5.12 Results provided by the simulated day-ahead market for the proposed 2030 Spanish scenario

5.5 Simulation Platform

Once the detailed models of the scenarios have been completed, the performance of the proposed TSO-DSO coordination schemes can be tested by means of tailored simulations. Having considered the complexity of the proposed scenarios (the necessity of modelling several devices based on different physical principles, the presence of both transmission and distribution networks, etc.), the project SmartNet [39, 40] has developed a dedicated software platform aimed at simulating the balancing and congestion management market interactions/clearing and their effects

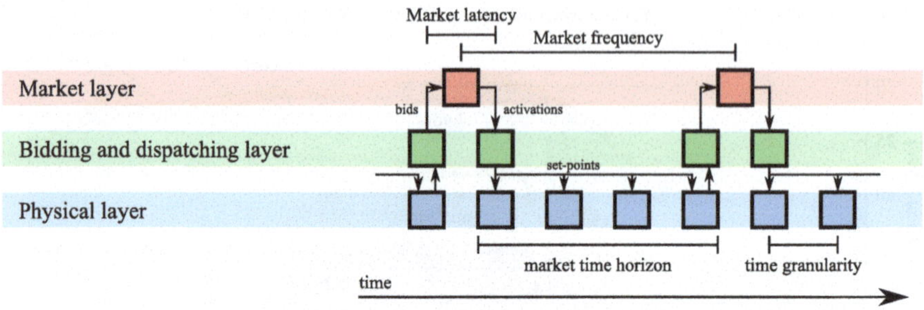

Fig. 5.13 Structure of the simulation platform highlighting the three simulated layers and the market timing parameters

on the physical system. Its structure can be seen as a combination of three layers (Fig. 5.13) which are sequentially processed:

- Bidding and dispatching layer (Sect. 5.5.1)
- Market layer (Sect. 5.5.2)
- Physical layer (Sect. 5.5.3)

All these layers deal with the flexibility of power units, which altogether compose the manual frequency restoration reserve (mFRR).

In addition to these main blocks, also information and communication technology (ICT) plays an important role. However, including the nonidealities of this dimension to the platform can unbearably increase the complexity of the simulation activity. However dedicated investigations (see Sect. 5.5.4) can be carried out in order to evaluate the adequacy of future (2030) ICT in the exploitation of the tasks in which system actors are involved.

5.5.1 Bidding and Dispatching Layer

The connection between the physical components of the system and the market is carried out by the bidding and dispatching layer. Two main procedures are processed in order to interact with the other two layers (Fig. 5.14):

- The bidding step consists of reading the availability of physical units in providing flexibility to the system. Depending on the technology, different costs have to be faced by the flexibility provider and the role of the bidder consists of offering an adequate price for it to the market. Further optimizations can be performed when more devices are aggregated in a single flexibility block.
- Once the market has selected the optimal activations, they have to be translated in dispatching orders and addressed to the flexible units. In case these are

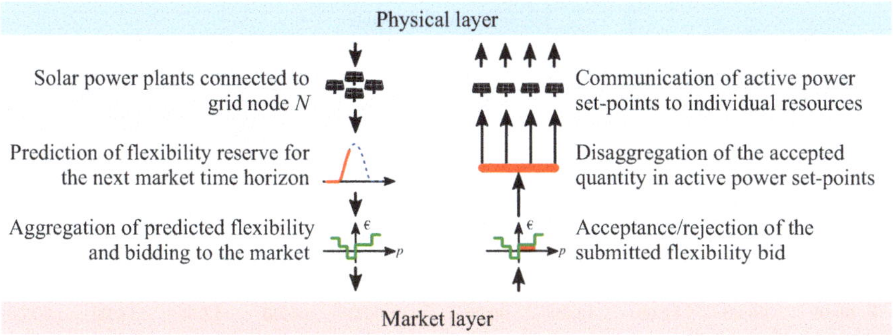

Fig. 5.14 Aggregation and disaggregation procedure, repeated for each node of the network (example with solar power plants connected to a single grid node N)

aggregated, (dis)aggregators need to appropriately subdivide the requested activation in a dispatching order for each controlled unit (on the basis of their actual availability).

One of the main challenges related to the bidding consists of predicting the availability of the resources over the (future) time instants for which the market is cleared. This aspect can result particularly critical for renewable-based resources, which can offer only an uncertain prediction of regulation reserve for the next time slots, and dispatching orders need, as far as possible, to take into account the impact of forecasting error.

Figure 5.15 reports the submitted and accepted mFRR for three different technologies, which are characterized by some peculiarities:

- *Bidding of curtailable generation*
  In this case, only downward flexibility is proposed as regulation reserve and it can be noticed that it has the characteristic profile of wind and solar generation (which is in line with the time availability of the renewables resources). In case of negative imbalance (positive mFRR required), this flexibility is purely used in order to manage congestions.
- *Bidding of thermostatically controlled loads*
  The regulation reserve provided by this kind of loads determines the presence of rebound effects, which can be easily noticed in Fig. 5.15. In fact, in case of negative imbalance, upward mFRR is activated, but this determines the activation of some (limited) downward reserve within the following time instants [41].
- *Bidding of electric vehicles*
  The amount flexibility depends on the actual availability of vehicles in the charging state. According to the assumed charging profiles (reported in Fig. 5.5), regulation reserve is not available early in the morning and in the late afternoon (when vehicles are on the road).

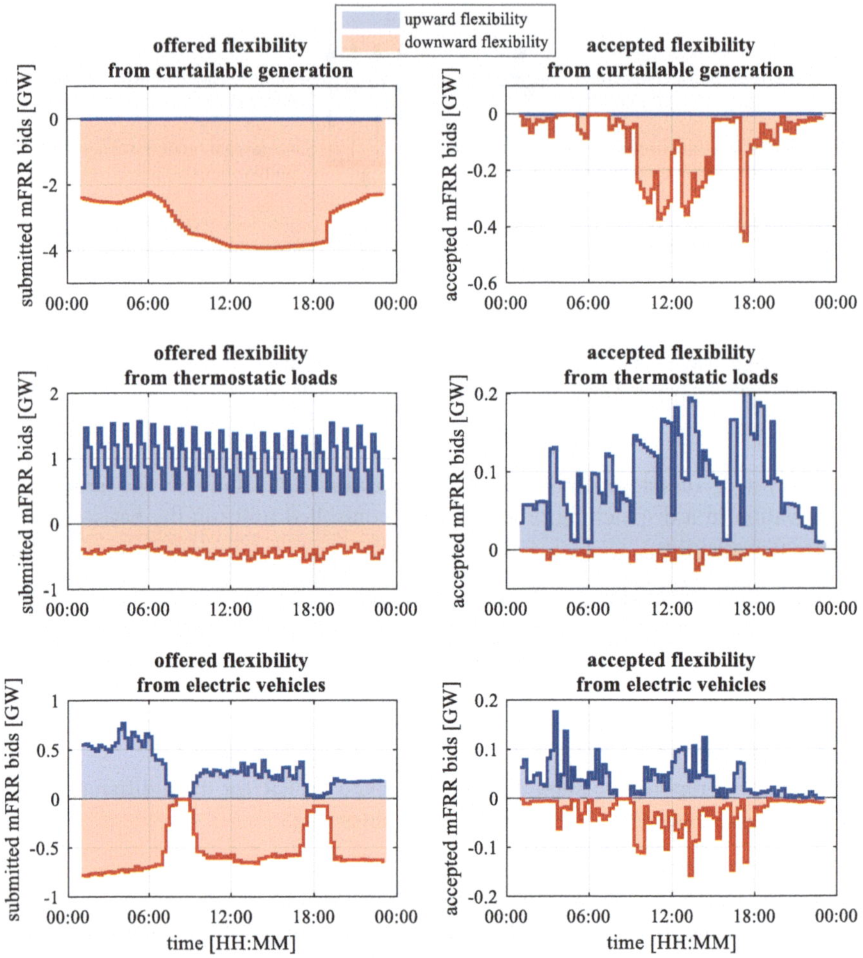

Fig. 5.15 Illustrative submitted and accepted mFRR bids for three different technologies: curtailable generation, thermostatically controlled loads and electric vehicles

5.5.2 *Market Layer*

This layer integrates the routines aimed at running the market clearing algorithms to solve system balancing and perform congestion management. As input it receives the bids generated by the bidding and dispatching layer and, having considered the network model/limitations, returns the optimal activations to solve the requested services.

One of the main novelties behind this layer consists of the management of the flexibility reserve provided by distribution resources (current power systems procure it mostly at transmission level) and the capability of considering distribution

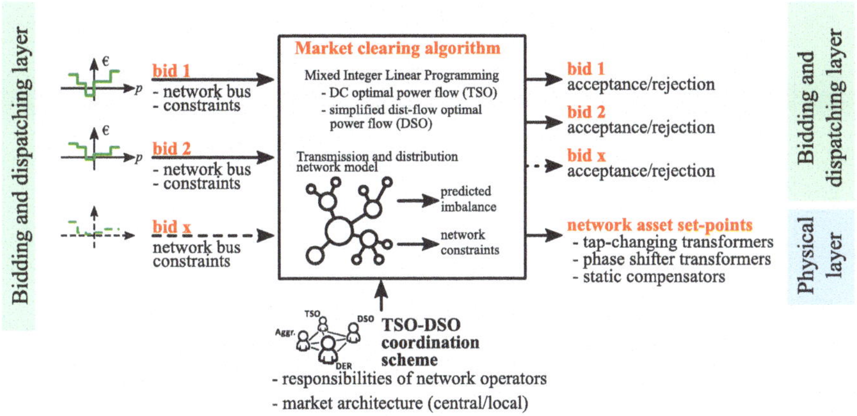

Fig. 5.16 Structure of the market layer, which takes as input the submitted bids, network model and TSO-DSO coordination schemes and returns as output the bids acceptance and the network asset set points

network constraints in its activation. In addition, since some distribution resources are characterized by non-conventional flexibility (i.e. time deferability, consumption profile alternation, storage capability), it also manages complex bids including integral constraints, alternative selections, etc. (see Chap. 3). This guarantees the possibility to distribution system operators to buy flexibility through the market in order to perform, depending on the TSO-DSO coordination scheme, local congestion management and balancing (Fig. 5.16).

Another important aspect is represented by the time dimension, which can be managed by the market clearing algorithm in a flexible way. In particular, the clearing routines can be run with the desired timing parameters in order to simulate different market dynamics (Fig. 5.13):

- *Market latency*, simulating the amount of time requested by the market clearing algorithm in order to process the input bids and to return the list of activations
- *Market frequency*, indicating how often the market process is launched
- *Market horizon*, representing the time range over which the activation of flexibility is requested

In addition to bidding devices, the market clearing algorithm has to take into account the flexibility of controllable network asset which can bring additional regulation margin for the available reserve. On-load tap changers of transformers, phase shifters and static compensators can be easily integrated within the clearing algorithm and used in order to enhance the participation of resources to the requested services (Fig. 5.16). Also, reactive power of flexible devices can provide significant contribution in terms of congestion management, especially for distribution grids where voltage issues can be often solved with an optimal management of resource power factor (Fig. 5.21).

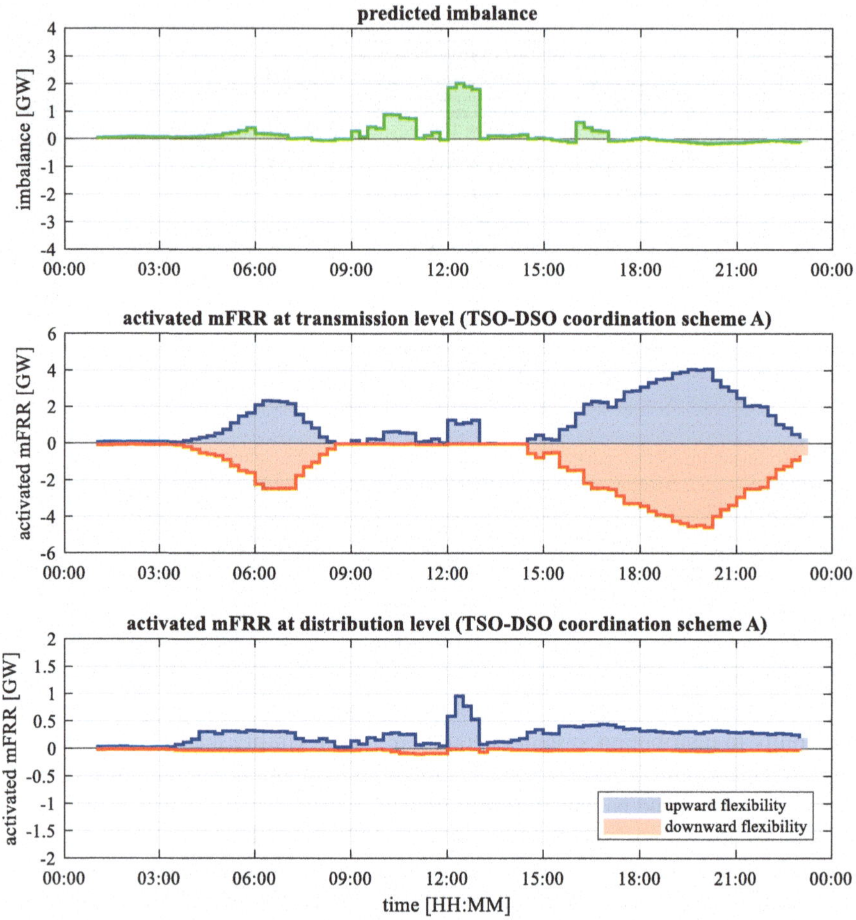

Fig. 5.17 mFRR activations resulting from the market layer during the simulation of a TSO-DSO coordination scheme in which DSO cannot buy flexibility for local congestion management

When multiple services are managed by the same market, regulation reserve can be activated simultaneously in two different directions. In fact, when network congestions occur, flexibility has to be activated in order to decrease grid loading but, at the same time, an opposite volume of reserve has to be managed in a non-congested portion of the network in order to rebalance the first one. This phenomenon can be noticed in Fig. 5.17, where the presence of congestions at transmission level is evident during the early morning and afternoon/evening.

From the same results it can be also noticed that upward distribution reserve results competitive with respect to transmission mFRR since many activations can be counted. On the contrary, almost no downward activations at distribution level are selected. However, allowing the DSO to buy flexibility for local services (congestion management) increases the usage of distribution downward reserve

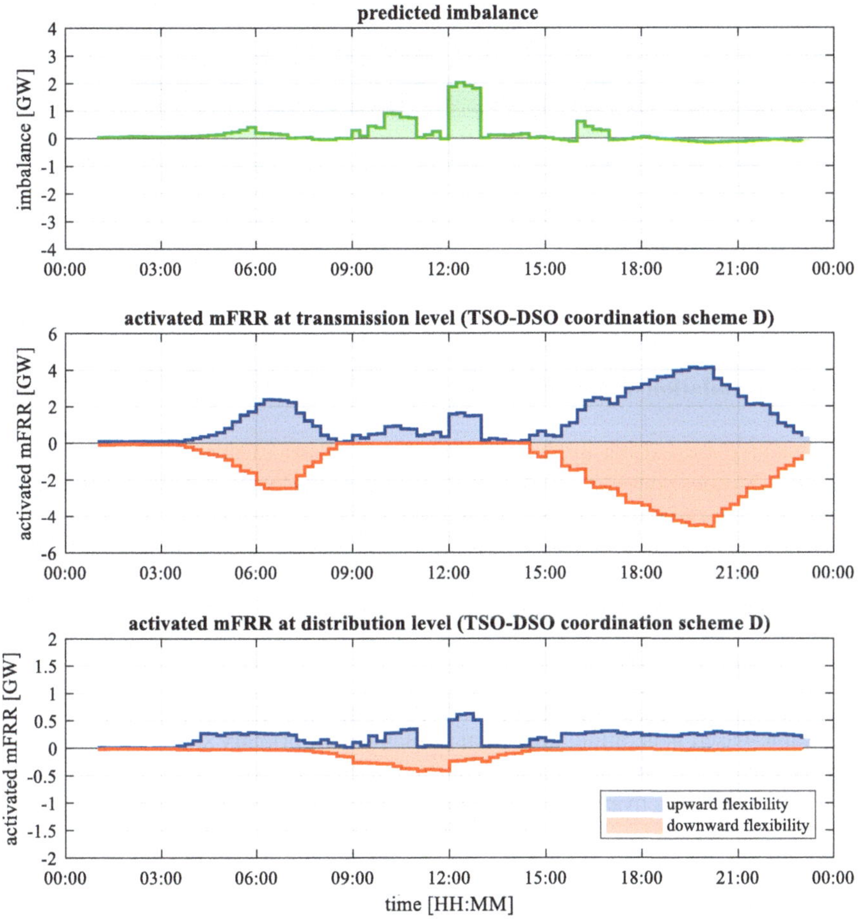

Fig. 5.18 mFRR activations resulting from the market layer during the simulation of a TSO-DSO coordination scheme in which DSO can buy flexibility for local congestion management

(Fig. 5.18). For this particular scenario, photovoltaic generation results to be the main cause of network congestions and this can be also noticed by looking at the shape of activated downward mFRR.

5.5.3 Physical Layer

Finally, all the decision-making process behind the bidding/aggregation is resulting in practical set points for the physical resources. This, in addition to have an impact on the internal states of the simulated devices, produces effects on the network that have to be carefully considered in order to analyse the performance of a specific TSO-DSO coordination scheme in the service management.

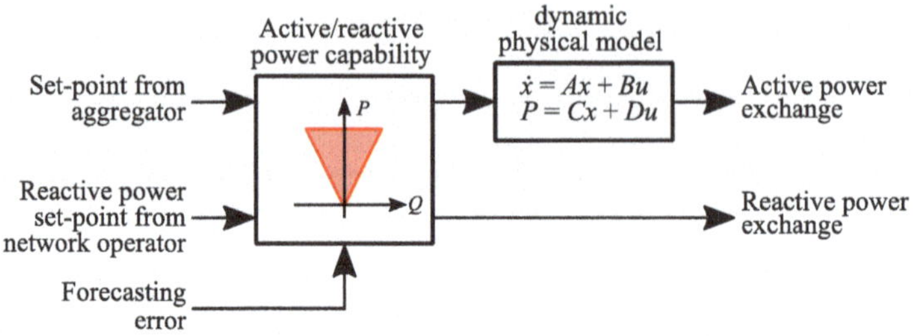

Fig. 5.19 Simulation of physical units taking into account the disaggregation set points, the possible interference of forecasting error on the power capability and internal dynamics

5.5.3.1 Simulation of Flexible Units

The first step carried out by the physical layer consists of determining the power output of physical devices connected to the network (Fig. 5.19). Of course, their behaviours depend on the set points requested by the bidding and dispatching layer and on the internal state of the devices. Depending on the technology, the adoption of zero- (renewable resources, atomic loads, conventional generators) and first-order (storage base devices, thermostatically controlled loads, CHPs) dynamic models can be considered a reasonable compromise between simplicity and accuracy [42].

Another important aspect to be considered is related to the forecasting error and its impact on the activations requested by disaggregators. In fact, especially for renewable-based generation, the actual availability of primary source (and flexibility) can differ from the one predicted during the bidding process. Of course, this is translated in a misactivation of the requested service which has to be managed by the physical layer.

In addition to active power, many simulated devices are also capable of providing reactive power flexibility. As discussed in the next subsection, network operators (also on the basis of market clearing results) manage this reserve for the voltage control of both transmission and distribution network. Of course, reactive power availability is dependent on the actual working point of the considered device and the related power capability.

5.5.3.2 Simulation of the Distribution and Transmission Network

Once the power output of each simulated device has been returned by the previous step, the network status can be easily computed. The simplest approach consists of running a conventional power flow, which returns the physical variables related to the grid (electrical voltage and currents). However, in order to have more realistic results, the following aspects should be considered [39]:

- The power exchanged by resources can cause unpredicted voltage/loading congestions that have to be promptly solved by the system operators by means of network asset management and, as last resort, re-dispatching of flexible devices.

- In addition to congestion management and balancing, other services are operated over the considered network. In particular, voltage control at transmission level has a significant impact on the power flows through the high-voltage network, sometimes bringing the necessity of tailored re-dispatching in case of congestions.

These points play a fundamental role in the simulation of the physical layer, since they keep the network in a stable operating area, even when the solution of the market and the requested activations cannot be practically actualized. The intervention of these additional measures is also expected to be dependent on the TSO-DSO coordination scheme. In particular, this is happening in market architectures where the DSO cannot buy flexibility. In this case, distribution network limits are disregarded by the market clearing algorithm and have to be manually managed by the DSO.

In order to actualize these measures, a possible strategy consists of running a dedicated optimal power flow (instead of a conventional one). The objective function can be constructed in order to prioritize actions on the network asset and to avoid, as far as possible, the use of flexibility products activated by the market (Fig. 5.20). In this order, the activations can be managed with the following priority scale:

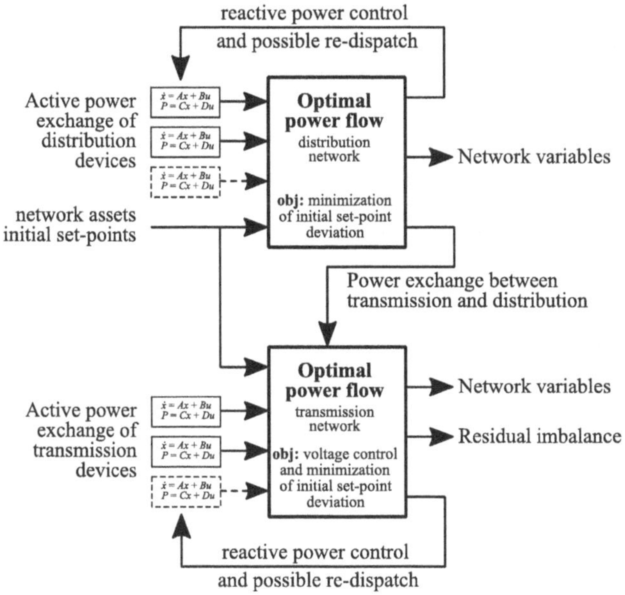

Fig. 5.20 Simulation of transmission and distribution network, including voltage control and possible re-dispatch (unwanted measures) in case of unexpected network congestions

1. On-load tap changing of transformers (OLTC), for the management of voltage congestions at distribution level
2. Reactive power control of flexible units, for voltage control at distribution and transmission level
3. Active power control of flexible units, for the management of grid overloading issues (in this case, the selected actions inevitably interfere with the market activations)

Thanks to the availability of distribution network asset (OLTC) and reactive power management of local resources (which is assumed to be not a remunerated service), many congestions can be solved without using market-activated reserve (mFRR in this case). Figure 5.21 reports the effects of DSO network management in terms of avoided voltage and loading congestions. In particular, it can be noticed how OLTC and reactive power flexibilities can dramatically decrease the voltage

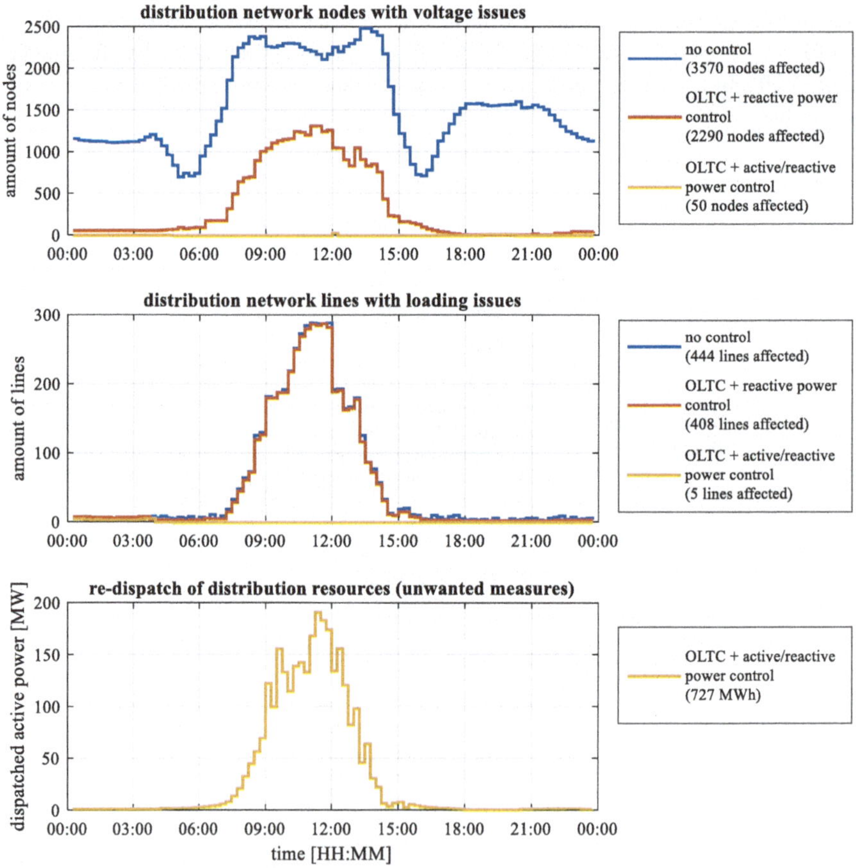

Fig. 5.21 Control of local asset and adoption of unwanted measures by the DSO in order to solve residual congestions (one scenario day of Italy)

issues. On the contrary, the only measure to effectively solve line overloading consists of active power regulation which, depending on the implemented TSO-DSO coordination scheme, corresponds to:

- *Activated mFRR*, when DSO can buy flexibility for local services by means of a dedicated local/centralized market
- *Unwanted measures*, when the DSO has to manually re-dispatch resources in order to promptly solve congestions, without accessing to the market

5.5.3.3 Management of Residual Imbalance (aFRR Activations)

After the application of the market/disaggregation set points to each resource, the correction of their actual power profile on the basis of forecasting error and unwanted measures activated by system operators, the network simulation returns a residual imbalance. In order to compensate it, real systems count on a dedicated power reserve: the automatic frequency restoration reserve (aFRR), which is aimed at maintaining the power frequency at 50 Hz and at restoring the power flows among neighbouring countries (i.e. cancelling internal imbalance). This is a crucial indicator of the performance of TSO-DSO coordination schemes, since it measures the ability of market/disaggregators in selecting/actuating mFRR for balancing and congestion management.

This typology of reserve is activated on the basis of a closed loop controller (which takes as error the power flow deviation on the country borders) and the participating resources are normally procured by the network operators through the balancing market. The selection of the resources typically happens within the same session used for the activation of mFRR, since there are some interferences between aFRR and mFRR. Nevertheless, the simulation of this process is adding significant complexity to the market clearing algorithm.

Possible simplifications can be assumed in order to realistically simulate aFRR activations without interfering with the three-layer simulation process described above. The approach adopted by the project SmartNet [43, 44] consists of an offline simulation of the aFRR procurement market (Fig. 5.22), based on the assumptions that:

- aFRR is provided by the same resources available on the mFRR market (Fig. 5.23).
- aFRR is equal to a fixed percentage of mFRR (since the simulated mFRR represents a real-time picture of the currently available flexibility), but more expensive (assumption made on the basis of actual statistics – see Fig. 5.23).
- The needed aFRR volume is normally dimensioned on the basis of the forecasting errors and typical residual imbalances. The simulated reserve is dimensioned by considering actual imbalance with some safety margins (Fig. 5.24).

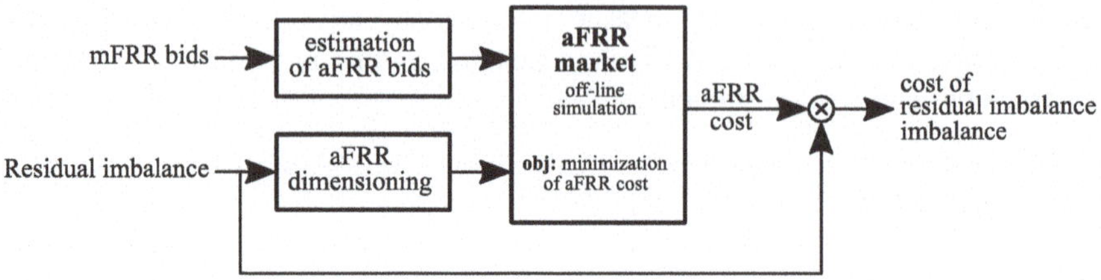

Fig. 5.22 Offline simulation aFRR market and calculation of residual imbalance cost

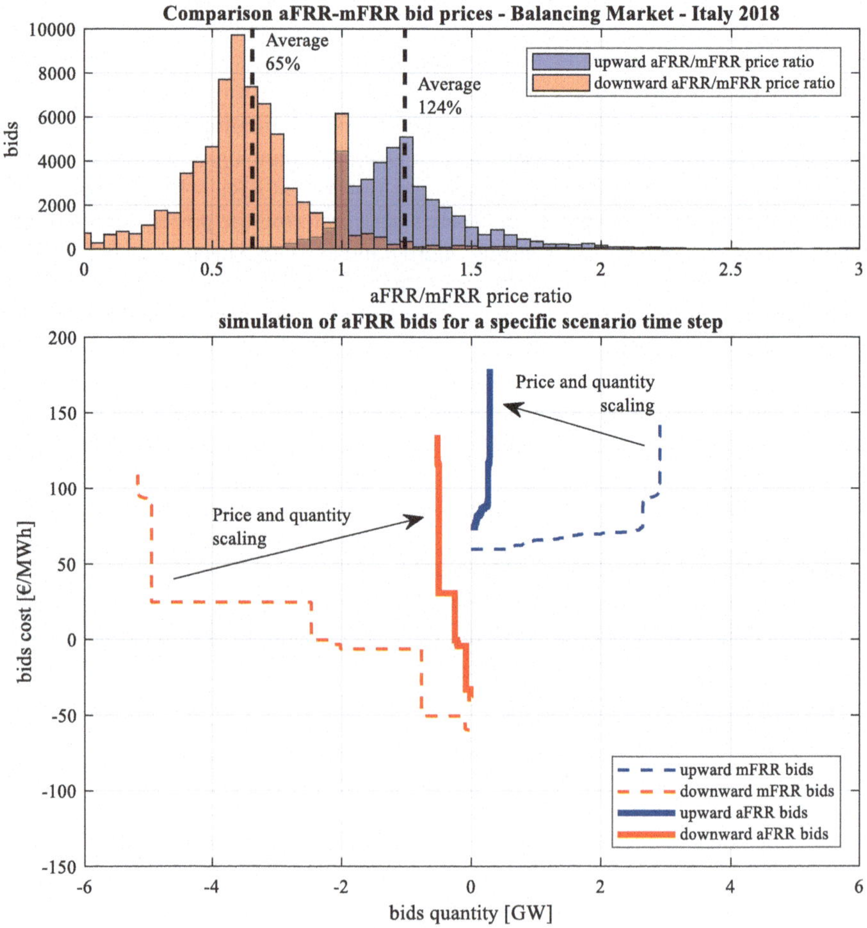

Fig. 5.23 Construction of the aFRR bidding curve on the basis of mFRR one for a specific scenario time step by scaling quantity (aFRR/mFRR = 10%) and costs

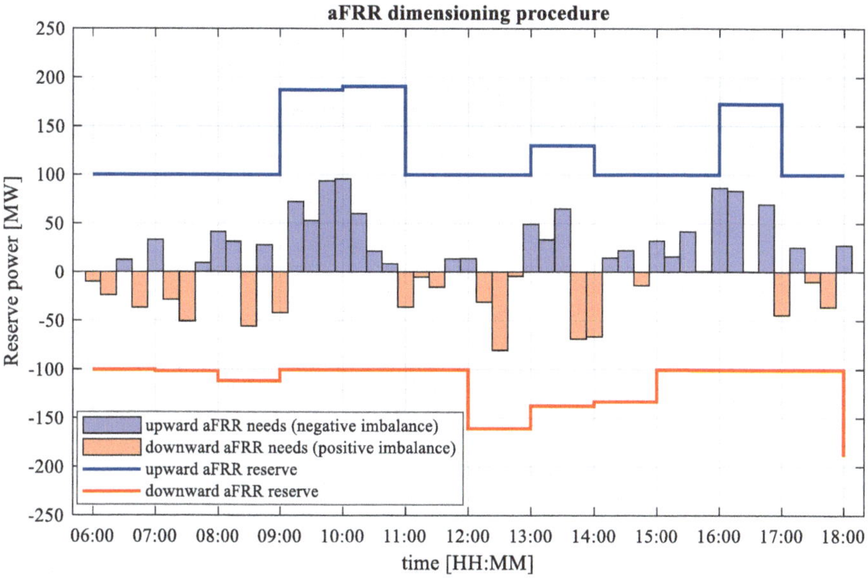

Fig. 5.24 Dimensioning of aFRR on the basis of the residual imbalance returned by the simulator (aFRR is assumed to be at least twice the resulting imbalance)

5.5.4 Communication Technology Layer

As anticipated above, the integration of information and communication technology models significantly increases the complexity of the simulation platform. However, simulations can be carried out having assumed that adequate technology for the requested services is in place within the considered scenarios.

This assumption can be easily verified by means of dedicated tests. In fact, having defined the communication technology requirements for the exploitation of specific ancillary services (see Chap. 4), its performance can be easily evaluated in controlled environments.

Thanks to the combination of the simulation platform described above and a laboratory facility, real equipment (aimed at supporting the provision of ancillary services) can be tested in realistic operating conditions. This is the concept of hardware-in-the-loop, which allows the inclusion of physical hardware in a simulated environment and the evaluation of its impact on the investigated services (e.g. impact of actual communication technology in a simulated 2030 scenario).

This analysis has been carried out by the project SmartNet [45, 46], which has considered (Fig. 5.25):

- A simplified 2030 scenario, capable of being simulated in real time, processed by means of the platform described above

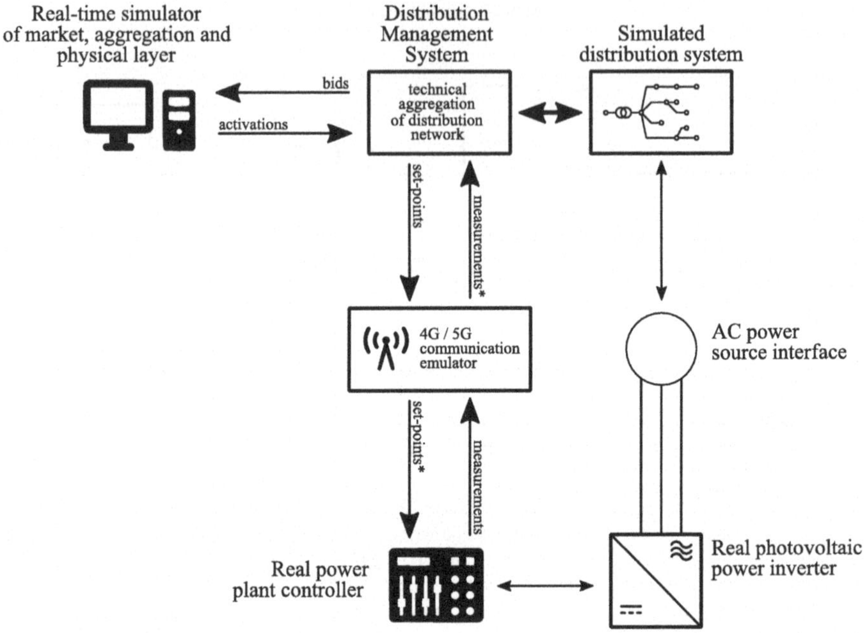

Fig. 5.25 Real-time simulation of 2030 scenario aimed at testing the performance of real equipment in presence of realistic communication systems

- A simulated distribution network and DSO management system, for which dedicated interfaces have been designed in order to interact with the simulated scenario
- A communication emulator, capable of reproducing the non-idealities of real communication technologies
- A real photovoltaic inverter (managed by a physical controller) and supplied by a synthetic photovoltaic power source

The advantage of this setup consists of the possibility of mimicking the non-ideal behaviour of actual communication technology. The communication emulator can be programmed in order to feature the same characteristics of the technology expected to be in operation in 2030 (4G/5G according to the analysis reported in Chap. 4) and the laboratory test results can be used in order to evaluate its performance.

The tests performed within the SmartNet project identify 4G/5G as a reliable technology for the provision of balancing and congestion management services (activation of both mFRR and aFRR).

5.6 Cost-Benefit Analysis

Once the simulation platform has been developed, the impact of the different TSO-DSO coordination schemes can be evaluated in a quantitative way over the hypothesized 2030 scenarios. As anticipated above, the investigated ancillary services consist of balancing and congestion management by means of manual frequency restoration reserve (mFRR). In particular, the simultaneous operation of these two services can be managed by the following four alternative coordination schemes:

CS A	*Centralized ancillary services market model* The DSO is not allowed to buy flexibility and market products can be acquired only by the TSO in order to perform balancing and congestion management at transmission level.
CS B	*Local ancillary services market model* The DSO has priority in order to buy flexibility for the management of distribution grid services (congestion management). The remaining flexibility (including the one located at distribution level) can be acquired by the TSO to activate balancing and congestion management.
CS C	*Shared balancing responsibility model* DSO and TSO manage their system separately, both performing balancing and congestion management with the resources available within their system.
CS D	*Common TSO-DSO ancillary services market model* DSO and TSO access to the same market to buy flexibility aimed at solving congestions and balancing. As for CS B, the DSO is responsible of the only congestion management at distribution level, but the flexibility is acquired within the same market used by the TSO for other services.

Chapter 2 foresees also a fifth possible TSO-DSO interaction (integrated flexibility market model) which is an extension of CS D where also non-regulated market parties can have access to the market under the same conditions as network operators. This coordination scheme is particularly difficult to simulate, especially for the crucial management of priorities that some services (e.g. congestion management) have among others (e.g. balancing).

The analysis of the mFRR activations related to these two services allows the comparison of the performance among the considered coordination schemes and the evaluation of the best trade-off between the added complexity brought by the implementation of complex market architectures and the lower system management cost. According to [43, 44], these aspects can be evaluated by comparing the following cost figures:

- Cost of activated mFRR
- Cost of activated aFRR
- Cost of unwanted measures
- Cost of information and communication technology

5.6.1 Cost of Activated Manual Frequency Restoration Reserve

The main result of the market clearing algorithm is represented by the mFRR activations aimed at compensating the predicted imbalance and solving the expected congestions. The selected resources depend on the simulated TSO-DSO coordination scheme and this has an impact on the cost. In particular, increasing the complexity of the market architecture (by adding constraints and services to the objective function) produces effects on the total cost of activated mFRR. This can be easily seen by looking at the simulations results (Fig. 5.26) obtained for the Danish 2030 scenario described above and investigated by the project SmartNet [43]:

- CS A, having less constraints (distribution grid services are not managed by this market architecture), provides lower activation costs with respect to the other coordination schemes.
- The inclusion of distribution network limits increases the mFRR costs since further resources have to be activated for the services requested by the DSO. The costs also depend on the way these services are managed: a centralized management (CS D) brings to lower mFRR costs if compared to the same services procured through a local market (CS B).
- The addition of other network constraints (such as the fixed profile exchange between transmission and distribution networks – CS C) further increases the mFRR costs.

In order to distinguish the portion of mFRR used for balancing and congestion management services, an additional simulation can be carried out by removing all the network constraints. This corresponds to a situation in which only imbalance is affecting mFRR activations. Once the costs are compared with the ones returned by the previous simulations, the economic impact of congestion management can be quantified (Fig. 5.26).

5.6.2 Cost of Activated Automatic Frequency Restoration Reserve

As anticipated above, the mFRR activations are based on the predicted imbalance and congestions and, for this reason, the effectiveness of the market results are dependent on the forecasting error and the correct consideration of network constraints (which have an impact on the application of unwanted measures).

Having assumed that the considered TSO-DSO coordination schemes are tested on the same scenarios (for which the forecasting error is the same), the differences in terms of aFRR are driven by the ability of the market clearing algorithm in managing network constraints. This aspect is particularly noticeable in the SmartNet simulation results obtained for the Spanish case (Fig. 5.27), for which:

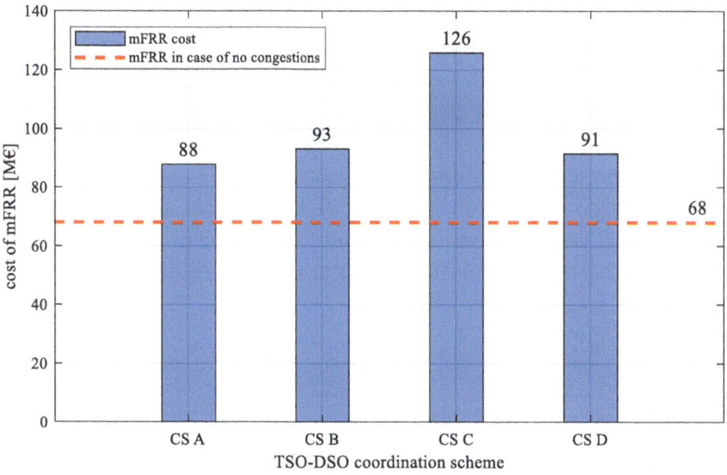

Fig. 5.26 Illustrative mFRR cost comparison among the investigated TSO-DSO coordination schemes (Danish 2030 scenario)

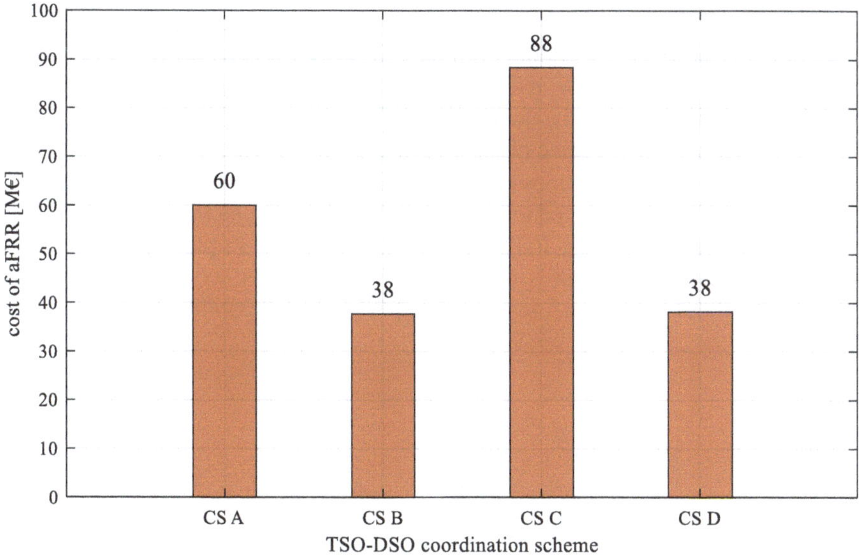

Fig. 5.27 Illustrative aFRR cost comparison among the investigated TSO-DSO coordination schemes (Spanish 2030 scenario)

- CS B and CS D feature the same aFRR activations since the related market clearing algorithm is based on a complete model of the network, capable of exhaustively managing all the grid constraints. In this case, the amount of activated aFRR corresponds to the imbalance caused by forecasting error.
- CS A has lower performance with respect to CS B/D and this is mainly due to the adopted market clearing algorithm, which activates mFRR resources regardless

of distribution network constraints. Blocked distribution resources and re-dispatching measures cause additional imbalance to be covered by further aFRR activations.
- CS C, in spite of the market clearing algorithm which implements the complete network model, includes the rigid balancing constraint of distribution network. According to the scenario assumptions described above, distribution systems can be rarely balanced because of lacking flexible resources at this level. This often produces market infeasibilities (also at transmission network, since TSO services can be provided by transmission resources only), resulting in high residual imbalances to be managed by aFRR.

5.6.3 Cost of Unwanted Measures

In case of unexpected congestions, network operators actuate measures in order to solve them promptly and to avoid the intervention of protection units (which determine power supply interruptions). Load/generation/storage re-dispatching and curtailment are the most common practices and they are operated without considering any economic merit order.

In order to monetize the adoption of these measures, assumptions on the way of remunerating them have to be made. One fair method for considering the provision of this service is assumed to be the payment of the actual cost of the used flexibility (which corresponds to the price of the mFRR bid – Fig. 5.28).

As anticipated in the previous section, unwanted measures (operated on the distribution network) are expected to be significant for CS A, and this is confirmed by the simulation results of the Italian scenario (Fig. 5.29). In particular, it can be noticed that unwanted measures at distribution level are not occurring when TSO-DSO coordination schemes foresee the acquisition of flexibility services to the DSO.

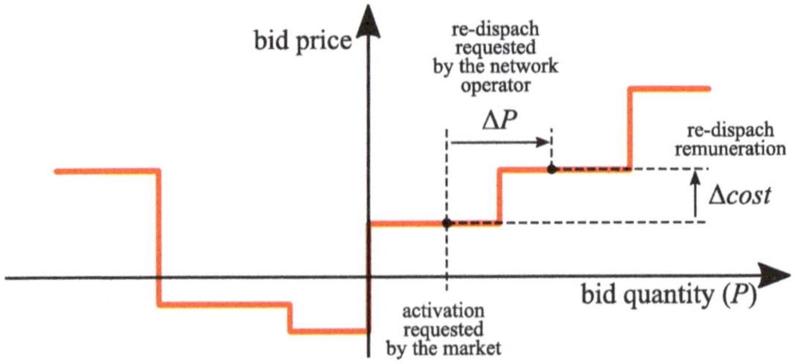

Fig. 5.28 Calculation of the cost afforded by network operators in case of unwanted measures

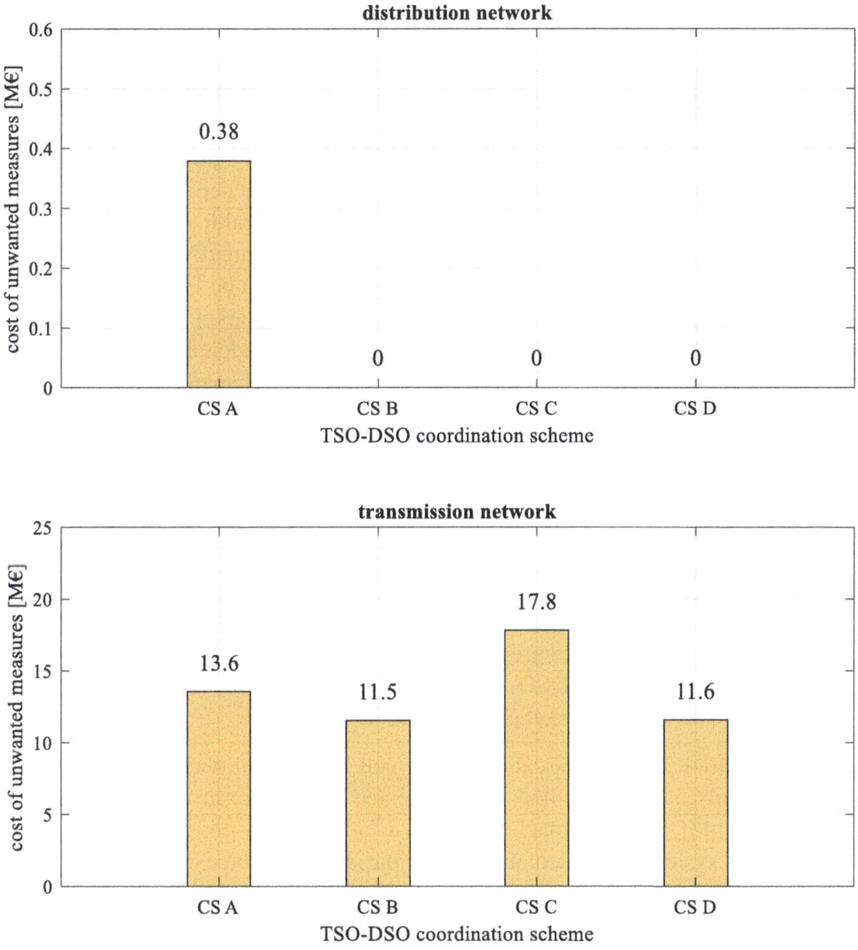

Fig. 5.29 Illustrative unwanted measure cost comparison among the investigated TSO-DSO coordination schemes (Italian 2030 scenario)

Re-dispatching is also occurring on the transmission network and unwanted measures have higher costs at high-voltage levels. For the Italian case, in particular, the lacking liquidity of resources of CS C is demonstrated by the high amount of re-dispatched resources.

Nevertheless, most of the time unwanted measures do not represent a significant system cost if compared to the ones to be attributed to the imbalance caused by them (and managed by aFRR – see Sect. 5.6.2).

5.6.4 Cost of Information and Communication Technology

Estimating the costs of information and communication technology (ICT) involves large uncertainties related to the technology development which is affected by the evolution of energy markets and grids up to 2030. In fact, aggregation and market clearing tasks require custom-made ICT which can vary a lot depending on few factors (e.g. market opportunity, contractual terms, financial health, etc.).

Looking at the investigations carried out by the project SmartNet [43, 47], TSO-DSO coordination scheme A can be reasonably assumed to be functional by 2030 since it corresponds to the natural evolution of the current market model. According to this, the last kilometre communication problem is considered solved in such a way that even the smallest flexible resource can be profitably aggregated to participate in TSO ancillary services markets. In particular, this assumption can be extended to any TSO-DSO coordination scheme since the communication requirements are exactly the same of CS A.

At this point, having also assumed that current TSO-DSO aggregator communication channels are adequate for the considered services, the main differences among coordination schemes substantially appears in terms of information technology (IT), especially:

- The extension needed in aggregating resources, which varies depending on the flexibility buyer (TSO or DSO)
- The complexity of the market clearing routines

The estimation of IT costs can be performed in different ways, and one of the most commonly adopted methods is represented by the Constructive Cost Model (COCOMO) [48]. This approach proposes a methodology to estimate the effort needed for IT system implementation on the basis of the source code lines.

Looking at the simulation platform described within the previous section, the code of the proposed bidding, market and dispatching blocks already integrates the main functionalities of fully industrialized solutions. This means that the IT costs can be estimated by considering:

- IT system implementation costs derived by the COCOMO method with the source code of the simulated bidding, market and dispatching blocks as input
- Costs for the adaptation to a high technology readiness level
- Extra costs due to increase in efforts in validating and testing fault-tolerant and dependable IT

These aspects have been considered within the project SmartNet [43, 47] and a detailed analysis has hypothesized the cost figures reported in Table 5.5.

At this point, having assumed a 10-year period with an interest rate of 5%, the resulting annuity costs for updating IT from CS A to other TSO-DSO coordination schemes are reported in Table 5.6. The indicated confidence intervals refer to the variable implementation effectiveness that can be achieved by local aggregators and/or markets.

Table 5.5 Estimated costs for updating information technology for the exploitation of TSO-DSO coordination schemes

IT update per system	Estimated cost (M€)
Aggregation of distribution resources for TSO services (CS A)	13.5
Update of aggregation from TSO services only to cover also DSO services (CS B, C, D)	10.6
Extension of centralized market for TSO services to distribution resources (CS A)	5.1
Development of local market for DSO congestion management services (CS B)	11.3
Development of local market for DSO congestion management and balancing services (CS C)	6.1
Update of the central market to consider both TSO and DSO services (CS D)	12.6

Table 5.6 Equivalent annuity costs in M€ with a 10-year period and 5% interest rate

CS A	CS A → CS B	CS A → CS C	CS A → CS D
4.59 ± 1.31	10.94 ± 2.39	9.17 ± 1.62	9.24 ± 1.02

The column related to CS A refers to the adaptation of market clearing and aggregation routines to enable distribution resource flexibility. Subsequent columns refer costs in updating CS A to other TSO-DSO coordination schemes

5.6.5 *Total Annual Costs*

Thanks to the cost figures reported in the previous subsections, the four considered TSO-DSO coordination schemes can be compared in terms of economic performance. Taking advantage of the results provided by the simulation platform described in Sect. 5.5, the peculiarities of each scenario can be analysed in detail and the most convenient TSO-DSO interaction identified for each of the investigated countries.

For the Italian scenario the results are reported in Fig. 5.30. According to them, CS B and CS D show the highest performance and this is due to the significant risk of distribution grid congestions. In fact, both these coordination schemes allow the DSO in exploiting flexibility services in order to solve local network problems without causing imbalance to the system. On the contrary, CS A returns higher costs especially for the adoption of (unbalanced) unwanted measures at distribution level which increase the necessity of aFRR activations (more expensive than mFRR – see Sect. 5.6.2).

For all the coordination schemes, even if mFRR is the most relevant cost figure, aFRR costs define their merit order in Italy. In particular, the optimal management of distribution grid (in conjunction with TSO services) is beneficial for the system and far larger than the additional costs brought by the more complex ICT infrastructure.

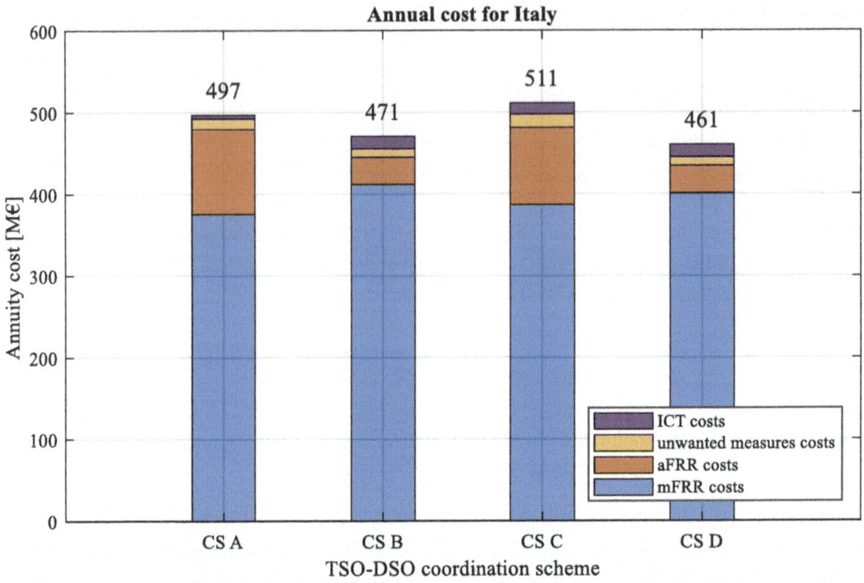

Fig. 5.30 Annual costs for balancing and congestion management services with the different TSO-DSO coordination schemes (Italy 2030)

Concerning the results of the simulated Danish scenario, opposite conclusions can be drawn. From Fig. 5.31 it can be noticed that CS A, CS B and CS D perform about the same costs in terms of mFRR and aFRR activations. This means that congestions at distribution level are not significant and allowing the DSO in acquiring flexibility on the market does not bring particular benefits to the system.

According to this, the merit order of the TSO-DSO coordination schemes in Denmark is mostly influenced by the ICT costs (in spite of being a small portion of the total costs) and CS B and CS D result to be less efficient than CS A.

Spain, instead, represents the middle ground with respect to the Italian and Danish case. In fact, the considered scenario hypothesized significant congestions at distribution level but, according to the results reported in Fig. 5.32, the benefits brought by a more efficient management of the system do not overtake the additional ICT costs and CS B/D result to be slightly less efficient than CS A.

For all the considered scenarios, it can be noticed that CS C is always the less beneficial TSO-DSO coordination scheme. This is mainly due to the poor liquidity of distribution resources for the management of local balancing services (the DSO is requested to balance the distribution network when CS C is implemented) and to the non-accessibility of TSO to flexibilities available at distribution level. From the analysis of the simulation results (the ones provided by the project SmartNet [23]) CS C frequently resulted unable to effectively rebalance all the distribution networks, bringing the necessity of high aFRR activations.

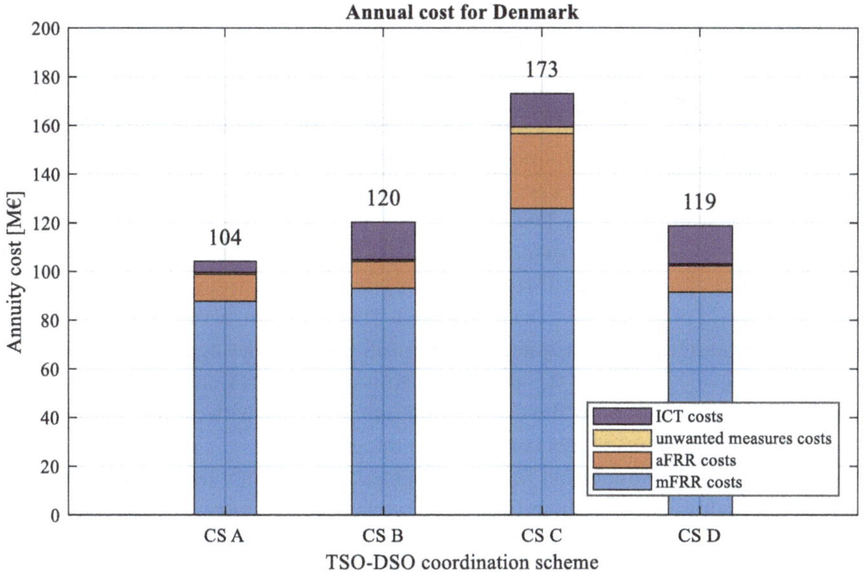

Fig. 5.31 Annual costs for balancing and congestion management services with the different TSO-DSO coordination schemes (Denmark 2030)

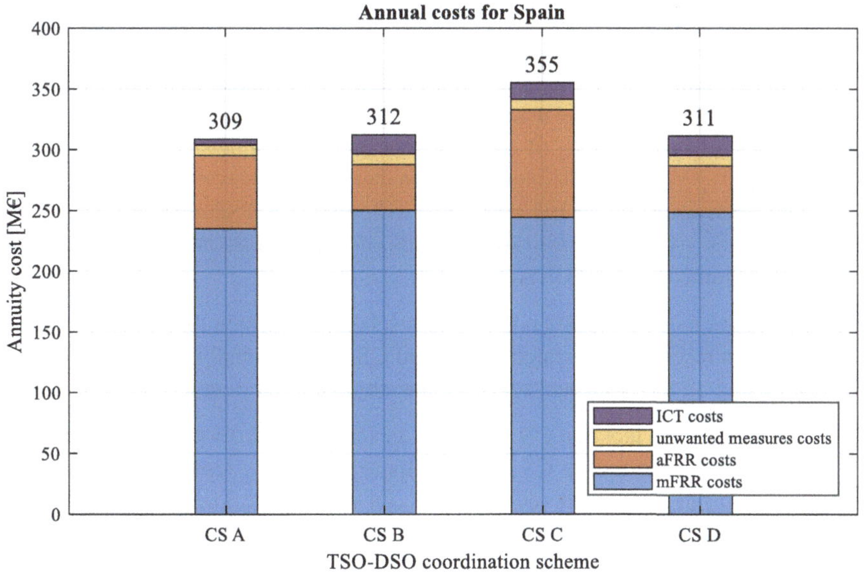

Fig. 5.32 Annual costs for balancing and congestion management services with the different TSO-DSO coordination schemes (Spain 2030)

5.7 Conclusions

Thanks to the data sources reported, a 2030 scenario has been defined for three countries: Denmark, Italy and Spain. For each country, different peculiarities have been highlighted especially pointing out how the distribution system is expected to behave differently in terms of congestions to be solved and, consequently, flexibility services to be acquired by the DSO.

In order to evaluate the performance of the proposed TSO-DSO coordination schemes when applied to the three scenarios, indications on how to build a dedicated simulation platform have been provided. This platform, actually developed by the project SmartNet, demonstrated the possibility of simulating the power system at capillary level, including the entire transmission network and a large portion (medium-voltage section) of the distribution one, together with all the hundreds of thousands of flexible resources connected to the grid.

According to the simulation results presented in the previous section, different conclusions can be drawn depending on the considered country and, therefore, the scenario design assumptions. In particular:

(a) The economic performance of TSO-DSO coordination schemes is mostly affected by the level of flexibility services requested by the DSO.

- When congestions on distribution network are not frequent and they occur with a probability comparable to forecasting error, CS A results to be the best compromise between system management effectiveness and costs. According to Sect. 5.6.5, this is the case of the Danish scenario, where DSO do not require large flexibility volumes and the economic benefits achieved with CS B/D are lower than the ICT development costs.
- When higher volumes of regulation reserve are requested by the DSO in order to perform local congestion management, the benefits in terms of optimal activation of mFRR (achieved with complex TSO-DSO coordination schemes – CS B/D) overtake the additional ICT costs necessary to update the assumed existing CS A.

(b) The adoption of TSO-DSO coordination schemes in which the management of regulation reserve is separated (CS B and CS C) is generally less efficient than optimising all the requested services in a single market iteration (CS D):

- In spite of the fact that CS B returns an economic performance very similar to the one featured by CS D, it results to be slightly less efficient than the latter in all the considered scenarios. Of course, the lower benefits can be explained by noticing that CS B splits the same mFRR optimization problem solved by CS D, introducing potential suboptimal solutions.
- CS C, instead, is definitively the TSO-DSO coordination scheme with the lowest performance. This is due to the introduction of the fixed power exchange constraint between distribution networks and transmission constraints, which significantly limits the liquidity of distribution resources and may lead to market clearing issues.

- Practical experience shows that, in rare circumstances (i.e. in the presence of severe congestions at transmission level), the splitting of market clearing routines can bring economic benefits, preventing the spreading of high prices (due to illiquidity of local markets) among distribution and transmission systems.

Of course, the presented simulation results and cost figures should be not assigned to the considered countries indelibly. The hypothesized scenarios are subjected to large uncertainties and, in particular, the assumptions related to the distribution networks (for which poor data is available). In fact, the presented investigation highlighted how distribution systems significantly impact on the performance of the proposed TSO-DSO coordination schemes.

Noticeable uncertainty can be attributed to ICT costs too, which in some cases reversed the merit order of the coordination scheme performance (see the Danish and Spanish case in Sect. 5.6.5). In addition, the reported analysis did not consider the problem with the last kilometre communication (infrastructure responsible of connecting aggregators with even the smallest resources). It might be possible that the related costs turn out to be too large for a profitable aggregation business.

References

1. J. Merino, I. Gómez et al., Ancillary services provision by RES and DSM connected at distributiol level in the future power system. H2020 SmartNet Deliverable 1.1. [Online]. Available: http://smartnet-project.eu/. Accessed 28 June 2019
2. European Commission, EU Reference Scenario 2016 (2016). [Online]. Available: https://ec.europa.eu/energy/sites/ener/files/documents/20160712_Summary_Ref_scenario_MAIN_RESULTS%20(2)-web.pdf. Accessed 28 June 2019
3. European Commission (IEC), Electrical Energy Storage (2011). [Online]. Available: http://www.iec.ch/whitepaper/pdf/iecWP-energystorage-LR-en.pdf. Accessed 28 June 2019
4. F. Thies, J. Rosenow, The Potential of Demand Response in Europe (2017). [Online]. Available: https://www.raponline.org/wp-content/uploads/2017/11/rap_sedc_rosenow_thies_fsr_slides_2017_oct.pdf. Accessed 28 June 2019
5. e-Highway Project. [Online]. Available: http://www.e-highway2050.eu/e-highway2050/. Accessed 28 June 2019
6. Roadmap2050 EU, Power Perspectives 2030 – On the Road to a Decarbonised Power Sector (2011). [Online]. Available: https://www.roadmap2050.eu/attachments/files/PowerPerspectives2030_FullReport.pdf. Accessed 28 June 2019
7. ENTSO-E, ENTSO-E 2015 Scenario Outlook & Adequacy Forecast (2015). [Online]. Available: https://docstore.entsoe.eu/Documents/SDC%20documents/SOAF/150630_SOAF_2015_publication_wcover.pdf. Accessed 28 June 2019
8. European Commission, A Roadmap for Moving to a Competitive Low Carbon Economy in 2050 (2011). [Online]. Available: http://eur-lex.europa.eu/legal-content/EN/ALL/?uri=CELEX:52011DC0112. Accessed 28 June 2019
9. E. C. Foundation, Power Perspectives 2030 – On the Road to a Decarbonised Power Sector (2011). [Online]. Available: http://www.roadmap2050.eu/attachments/files/PowerPerspectives2030_FullReport.pdf. Accessed 28 June 2019
10. European Commission, 2050 Low-Carbon Economy. [Online]. Available: https://ec.europa.eu/clima/policies/strategies/2050_en#tab-0-0. Accessed 28 June 2019

11. T.D. Government, The Danish Climate Policy Plan. Towards a Low Carbon Society (2013). [Online]. Available: https://ens.dk/sites/ens.dk/files/Analyser/danishclimatepolicyplan_uk.pdf. Accessed 28 June 2019
12. smartEn Smart Energy Europe, Explicit Demand Response in Europe – Mapping the Markets (April 2017). [Online]. Available: http://www.smarten.eu/explicit-demand-response-in-europe-mapping-the-markets-2017/. Accessed 28 June 2019
13. European Commission, 2030 Framework for Climate & Energy. Outcome of the October 2014 European Council. [Online]. Available: https://ec.europa.eu/clima/sites/clima/files/strategies/2030/docs/2030_euco_conclusions_en.pdf. Accessed 28 June 2019
14. ENTSO-E, Ten Year Network and Development Plan (2016). [Online]. Available: http://tyndp.entsoe.eu/exec-report/. Accessed 28 June 2019
15. F. Lanati, A. Gelmini, M. Benini, M. Gallanti, A methodology to assess the impact of 2030 EU climate and energy targets on the national power systems: The Italian case. de 2015 12th International Conference on the European Energy Market (EEM), Lisbon, 2015
16. E. Gavaghan, EU 2030 objectives: Emissions vs. renewables. Environment & Energy Affairs (30 Jan 2014). [Online]. Available: http://www.europeanpublicaffairs.eu/eu-2030-objectives-emissions-vs-renewables/. Accessed 28 June 2019
17. ENTSO-E – European Network of Transmission System Operators for Electricity, ENTSO-E Transparency Platform. [Online]. Available: https://transparency.entsoe.eu/. Accessed 28 June 2019
18. Enerdata, Power Plant Tracker. [Online]. Available: https://www.enerdata.net/research/power-plant-database.html. Accessed 28 June 2019
19. Gestore dei Servizi Energetici GSE S.p.A., AtlaImpianti – Dati e Scenari. [Online]. Available: https://www.gse.it/dati-e-scenari/atlaimpianti. Accessed 28 June 2019
20. The Windpower – Wind Energy Market Intelligence. [Online]. https://www.thewindpower.net/. Accessed 28 June 2019
21. F.G. Montoya, M.J. Aguilera, F. Manzano-Agugliaro, Renewable energy production in Spain: A review. Renew. Sustain. Energy Rev. **3**, 509–531 (2014)
22. Gestore dei Servizi Energetici GSE S.p.A., Atlante impianti fotovoltaici in Italia. [Online]. Available: http://atlasole.gse.it/atlasole/. Accessed 28 June 2019
23. U.S. DOE, Office of Electricity and Sandia National Labs, United States Department of Energy Global Energy Storage Database (GESDB). [Online]. Available: https://www.energystorageexchange.org/. Accessed 28 June 2019
24. M. Rossi, H. Svendsen et al., Scenario set up and simulations results. H2020 SmartNet Deliverable 4.2. [Online]. Available: http://smartnet-project.eu/. Accessed 28 June 2019
25. European Commission, Eurostat. [Online]. Available: https://ec.europa.eu/eurostat/data/database. Accessed 28 June 2019
26. B. Grinden, N. Feilberg, Analysis of Monitoring Campaign in Europe (REMODECE D10). [Online]. Available: https://remodece.isr.uc.pt/. Accessed 28 June 2019
27. M. Pipattanasomporn, M. Kuzlu, S. Rahman, Y. Teklu, Load profiles of selected major household appliances and their demand response opportunities. IEEE Trans. Smart Grid **5**(2), 742–750 (2014)
28. R. Stamminger et al., Synergy Potential of Smart Appliances (Smart-A D3.2). [Online]. Available: https://ec.europa.eu/energy/intelligent/projects/en/projects/smart-0. Accessed 28 June 2019
29. ENTSO-E – European Network of Transmission System Operators for Electricity, ENTSO-E: Map of the Transmission Grids. [Online]. Available: https://www.entsoe.eu/data/map/. Accessed 28 June 2019
30. Energinet, Transmission System Data. [Online]. Available: https://en.energinet.dk/Electricity/Energy-data/System-data. Accessed 28 June 2019
31. Ministero dell'Ambiente e della Tutela del Territorio e del Mare, Rete elettrica – stazioni e centrali. [Online]. Available: https://va.minambiente.it/it-IT/DatiEStrumenti/MetadatoStrato/50f26f40-1f6f-0b49-8c45-0f3ba5d62e52. Accessed 28 June 2019

32. ENTSO-E – European Network of Transmission System Operators for Electricity, TYNDP 2018 – Europe's Network Development Plan to 2025, 2030 and 2040. [Online]. Available: https://tyndp.entsoe.eu/. Accessed 28 June 2019
33. Energinet, System Plan 2018 – Electricity and Gas in Denmark. [Online]. Available: https://en.energinet.dk/About-our-reports/Reports/System-Plan-2018. Accessed 28 June 2019
34. Terna, The National Electricity Transmission Grid Development Plan – Operation and objectives of the Terna Development Plan. [Online]. Available: https://www.terna.it/en-gb/sistemaelettrico/pianodisviluppodellarete.aspx. Accessed 28 June 2019
35. F. Pilo, G. Pisano, S. Scalari, D. Dal Canto, A. Testa, R. Langella, R. Caldon, R. Turri, ATLANTIDE – Digital archive of the Italian electric distribution reference networks. CIRED 2012, Lisbon, May 2012
36. G. Viganò, M. Rossi, D. Moneta, Generation of artificial distribution networks for the evaluation of advanced control solutions, in *2017 AEIT International Annual Conference*, Cagliari, 2017, pp. 1–6
37. H.G. Svendsen, O.C. Spro, PowerGAMA: A new simplified modelling approach for analyses of large interconnected power systems, applied to a 2030 Western Mediterranean case study. J. Renew. Sustain. Energy **8**, 055501 (2016)
38. H.G. Svendsen, L. Warland, M. Korpås, D. Huertas-Hernado, J. Völker, Report describing the power market model, data requirements and results from analysis of initial grid design. OffshoreGrid project (2010). [Online]. Available: http://www.offshoregrid.eu/index.php/results. Accessed 28 June 2019
39. G. Viganò et al., SmartNet simulation platform. H2020 SmartNet Deliverable 4.1. [Online]. Available: http://smartnet-project.eu/. Accessed 28 June 2019
40. M. Rossi, G. Viganò, G. Migliavacca, Y. Vardanyan, R. Ebrahimi, G. Leclercq, P. Sels, M. Pavesi, T. Gueuning, J. Jimeno, N. Ruiz, G. Howorth, J. Camargo, C. Hermans, F. Spiessen, H. Svendsen, Testing TSO-DSO interaction schemes for the participation of distribution energy resources in the balancing market: The SmartNet simulator. *CIRED*, 2019, paper n° 1158
41. J. Jimeno, N. Ruiz, C. Madina, Aggregation of thermostatically controlled loads for flexibility markets. *CIRED*, 2019, paper n° 1502
42. J. Le Baut, G. Leclercq, G. Viganò et al., Characterization of flexibility resources and distribution networks. H2020 SmartNet Deliverable 1.2. [Online]. Available: http://smartnet-project.eu/. Accessed 28 June 2019
43. C. Madina et al., Cost-benefit analysis of the selected national cases. H2020 SmartNet Deliverable 4.3 (2019). [Online]. Available: http://smartnet-project.eu/. Accessed 28 June 2019
44. C. Madina, S. Riaño, I. Gómez, P. Kuusela, H. Aghaie, J. Jimeno, N. Ruiz, M. Rossi, G. Migliavacca, Cost benefit analysis of TSO-DSO coordination schemes to operate flexibility markets. *CIRED*, 2019, paper n° 1632
45. F. Pröstl Andrén, Strasser T., C. Seitl, G. Viganò, M. Rossi et al., Lab-environment set-up and simulations. H2020 SmartNet Deliverable 4.4 (2019). [Online]. Available: http://smartnet-project.eu/. Accessed 28 June 2019
46. F. Pröstl Andrén, T. Strasser, J. Le Baut, M. Rossi, G. Viganò, G. Della Croce, S. Horsman-Heimo, A. G. Azar, A. Ibañez, Validating coordination schemes between transmission and distribution system operators using a laboratory-based approach. IEEE PES Powertech (2019)
47. P. Kuusela, P. Koponen, I. Kockar, H. Xu, An ICT cost comparison of different market structures for distributed ancillary services. *CIRED*, 2019, paper n° 827
48. B. Boehm, *Software Engineering Economics* (Prentice Hall, Englewood Cliffs, 2981)

Chapter 6
Technologies and Protocols: The Experience of the Three SmartNet Pilots

Carlos Madina, Joseba Jimeno, Luca Ortolano, Margherita Palleschi, Razgar Ebrahimy, Henrik Madsen, Miguel Pardo, and Cristina Corchero

6.1 Introduction

The coordination between transmission system operators (TSOs) and distribution system operators (DSOs) is an important topic to be developed for the future integration of the electricity production from renewable energy sources (RES) and other types of distributed energy resources (DER) into power systems, as discussed along this book.

Since there are few real-life experiences in the application of the concepts described in previous chapters, the deployment of technological pilots is very important to test and demonstrate the technical feasibility of the different coordination schemes.

In previous chapters, different options for coordinating TSOs and DSOs have been described, and the process to perform a cost-benefit analysis (CBA) has been discussed in detail. Based on such CBA, regulatory authorities and public bodies can identify the most suitable coordination scheme to be implemented within one country under some given conditions, defined by the generation and demand

C. Madina (✉) · J. Jimeno
TECNALIA, Derio, Spain
e-mail: carlos.madina@tecnalia.com

L. Ortolano · M. Palleschi
Terna, S.p.A., Rome, Italy

R. Ebrahimy · H. Madsen
Danmarks Tekniske Universitet, Lyngby, Denmark

M. Pardo
e-distribución, Barcelona, Spain

C. Corchero
IREC Institut de Recerca en Energia de Catalunya, Barcelona, Spain

G. Migliavacca (ed.), *TSO-DSO Interactions and Ancillary Services in Electricity Transmission and Distribution Networks*,
https://doi.org/10.1007/978-3-030-29203-4_6

scenario. However, bringing the coordination schemes to real life may result in some implementation difficulties which cannot be anticipated by the scenario analysis and CBA, and that is a second main reason for implementing real-life technological pilots.

Whenever possible, more than one pilot should be deployed, so that each of them can focus on different parts of the TSO-DSO coordination value chain. In that sense, it is important to demonstrate different potential TSO-DSO coordination schemes, so that issues arising from each of them can be identified. Moreover, it is also important to demonstrate different types of DER, so that their flexibilities can be better assessed and the advantages and disadvantages for real-life implementation can be properly identified and addressed. As a third complementarity aspect, having different technological pilots allows for focusing on different parts of the value chain, so that one of them may take the vision of the TSO or DSO, while another one can focus on the needs of the aggregator or DER owners.

Within the SmartNet project, three complementary pilots were implemented and are described in sections below. As the reader will discover, they are complementary in terms of geographical implementations, but also in the different aspects described above.

The first pilot was deployed in an area with high penetration of RES in Northern Italy. In this mountainous area, there are many run-of-river hydropower plants and not a big demand, resulting in a situation with reverse power flow, i.e. power going from distribution up to transmission, for most of the year. One of the main tasks is on the aggregation of load and generation information by the DSO, in order to obtain both real-time information and a better forecast of the grid conditions in upcoming periods. This information is very useful for the TSO to anticipate (and avoid) problems in the transmission network and to know in real time the available flexibility sources that can be used by the TSO respecting the constraints of the distribution grid. Moreover, it can also be used to estimate the flexibility that DER could provide for voltage control or system balancing managed in a centralized scheme by the TSO.

The second pilot was installed in Denmark, with the objective to demonstrate the potential of exploiting the flexibility of indoor swimming pools by using price signals. The thermostats of swimming pools can be controlled to react to different price levels, in order to consume more or less energy. Therefore, an aggregator can calculate the flexibility function of swimming pools, i.e. how much flexibility the swimming pools will provide for different price levels, and broadcast the appropriate price signals to obtain a given flexibility level. Then, such price-flexibility function can be used by the aggregator to bid into the markets for ancillary services. For the purpose of the pilot, the coordination scheme in which both the TSO and the DSO post their balancing and congestion management needs was selected.

The third pilot was implemented in Spain and aimed at demonstrating the technical feasibility of creating a new market, managed by the DSO, to procure ancillary services from units connected at distribution level. In this case, the shared balancing responsibility coordination scheme was selected, so that the DSO is responsible, not only to solve congestions at distribution level but also to maintain a scheduled power exchange profile at the TSO-DSO interconnection, in order to reduce the imbalances to be solved by the TSO. The Spanish pilot exploited the flexibility available in radio base stations for mobile phone communications, leveraging the backup batteries installed to maintain the communications service in case of a blackout, but which are almost never used. Therefore, aggregators calculate the available flexibility in base stations and bid it into the local market, so that the DSO can solve congestions and maintain the agreed profile in the TSO-DSO interconnection.

Table 6.1 summarizes the complementarity of the three pilots.

Being in the vanguard of technology implementation, these pilots uncovered a number of issues, ranging from regulatory (such as impeding DER to participate in the markets for ancillary services operated by the TSO or having different metering requirements depending on the contracted consumption power) to technical (such as low mobile phone connectivity in remote rural areas or faulty backup batteries, which, fortunately, never had to provide backup power until the pilot started the testing phase) and even practical barriers (e.g. radio base stations are located in the roofs of residential buildings, so replacing their cabinets requires obtaining permission from landlords but also from municipalities, as they must be uploaded by huge cranes located in the streets).

Table 6.1 Complementarity of technological pilots

	Pilot A	Pilot B	Pilot C
Country	Italy	Denmark	Spain
Coordination scheme	Centralized ancillary services market	Common TSO-DSO ancillary services market	Shared balancing responsibility
Services to be gathered by TSO-DSO	Aggregation of information for TSO Voltage control for TSO Frequency control for TSO	DSO congestion management Frequency control for TSO	DSO congestion management Frequency control for DSO
DER providing flexibility	Run-of-river hydropower plants	Impulsion pumps for hot water for indoor swimming pools in rental houses	Backup batteries for radio base stations used in mobile phone communications
Main focus of the pilot	TSO-DSO communication TSO control Assessment of DER capability to participate in markets	Price signals from aggregators to obtain DER flexibility Communication chain from market to DER through aggregators	Monitoring of distribution network Creation and operation of local flexibility markets Assessment of base station capability to provide services for grid support

6.2 TSO Focus: Increased Observability Over the Distribution Network

6.2.1 *Challenges Arising from High DER Penetration*

In order to lead the energy transition, the European Union has adopted a policy that aims to encourage the development of new and renewable forms of energy and the replacement of fossil fuels.

In this context, Italy has faced an important growth of RES penetration, mostly wind and photovoltaic (PV). As shown in Fig. 6.1, about 26 GW of RES have been installed (6.4 GW of wind farms and 19.4 GW of PV panels) since 2008, thanks to the support of the government incentives. Already reached the 2020 objective, the targets set by the National Energy Strategy aim at reaching a level of 28% of gross energy consumption at 2030 (in particular a 55% share of RES in electricity consumption) and for the complete phase out from coal by 2025.

There are two main characteristics of these forms of energy generation that can affect the management of the electrical system. On one hand, they have a variable behaviour that depends on aleatory primary energy sources (wind, sun, water). On the other hand, small-sized generators, mainly PV and run-of-river hydropower plants, are usually connected to the distribution network. As a result, the growth of RES penetration is closely linked to the spread of the distributed generation.

The consequence is that the energy framework is moving from a power generation structure mainly characterized by few big traditional plants connected to the high-voltage (HV) transmission grid and controlled directly by the TSO to a generation park composed by numerous plants connected to medium-voltage (MV) and low-voltage (LV) grids. This transformation of the generation mix leads to three main critical issues for the management of the power system: reverse flow, challenges for frequency and voltage regulation and reduction of available reserve.

In the past, power systems were designed by considering distribution grids traditionally as "passive" networks. Under this paradigm, the active power flows were

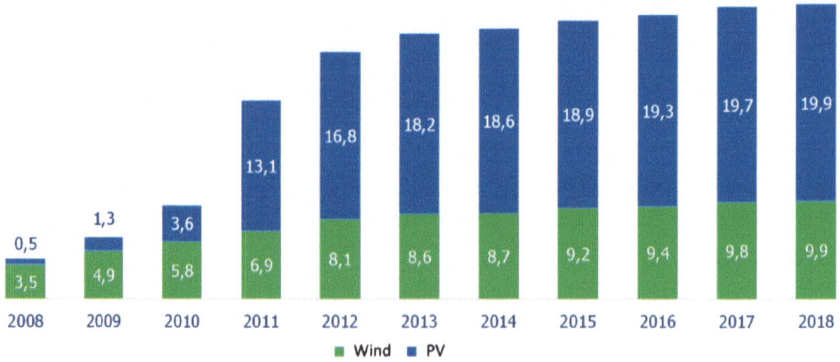

Fig. 6.1 Wind and PV capacity installed in Italy (GW), 2008–2018 [1]

Fig. 6.2 Example of reverse flow at the HV/MV transformer of a primary substation

unidirectional, from the HV grid, where generation plants were connected, to the distribution grid, mainly composed by loads. Due to the spread of DER connected at lower voltage levels of the network, the new configuration may create reverse power flows, i.e. when the distributed generation exceeds the local consumption behind the same substation, the local oversupply at the distribution grid results in active power rising from lower (distribution grid) to upper voltage levels of the grid (transmission grid), as represented in Fig. 6.2.[1]

The bidirectional nature of power flows influences the management and operation of the transmission grid. For instance, the reduction of the active power flow at the TSO-DSO interconnection point causes a voltage increase in the HV grid, because the power line produces reactive power when the active power along a feeder is below the nominal value. Therefore, there is an increasing need of sources to control and regulate the grid voltage.

Besides, the monitoring and automation devices installed at the HV side of the primary substation, particularly in radial networks, are designed for unidirectional power flows, and the bidirectionality of the power flow can affect their efficiency when monitoring and operating the network.

The reverse power flow has a third impact in grid operation under emergency conditions. In case of under-frequency, the defence plan prescribes the load curtailment to restore the power balance. However, due to the presence of generation connected at the distribution level, the net load disconnected is lower than in the past, causing a reduction of the selectivity and efficiency of the action.

In addition to the reverse power flow, another consequence of the high penetration of non-programmable, distributed generation implies issues related to the frequency and voltage profiles regulations. The non-programmable nature of RES implies the unpredictability of the production and the impossibility to ensure a fixed power exchange with the power system. For this reason, the current regulation does not require these power plants to provide ancillary services (either to ensure system balance or for reactive power control). As a result, thermoelectric plants must be

[1]The control system of Italian TSO uses a passive sign convention, and the power value is negative when active power is supplied to the grid.

kept in operation, even at off-peak times, to provide ancillary services and ensure the availability of the required reserves.

Linked to this unpredictability, the spread of RES leads to the reduction of the active power reserve: the downward reserve (i.e. the reduction of the production) decreases during daylight hours when the PV generation is high, because PV is currently not controllable by the TSO when connected at the distribution level. Moreover, the upward reserve (i.e. the increase of the production) is reduced during drought/cold periods due to the reduction of hydroelectric generation and the increase of the peak load demand.

All these exemplifying aspects have to be considered by the TSO in the management of the power system in future scenarios, which is expected to be characterized by a high RES penetration. For this reason, the Italian TSO is launching numerous initiatives to enable the integration of RES in all the phases of the management of the grid: in forecast calculation, during the planning phase, in real-time operation, in the provision of ancillary services, in the defence plan, etc.

6.2.2 The Importance of Grid Monitoring

In order to enable distributed generation to become an active player in the electricity system, the monitoring of the whole grid (including lower voltage levels) is a prerequisite to give TSO the needed perception of the energy mix underlying the primary substation, the geographic allocation of the generation, the actual energy consumption of the load and the response of the power plants in the provision of the services.

Indeed, the TSO needs a continuous monitoring and a transparent access to the MV and LV levels of the grid, to ensure a safe and reliable management of these resources, particularly to handle non-programmable sources.

In addition to allowing the participation of distributed generation in the provision of ancillary services, the knowledge of the real-time conditions of the whole grid allows the TSO to improve the efficiency of the tool used for grid management. For instance, the grid calculations currently use a gross estimate based on grid models and on the installed generation capacity, i.e. for state estimation and for static and dynamic simulation (both online and offline) to identify critical constrains. Furthermore, it could enhance the defence system, by adapting the protection scheme of the system due to a better assessment of the N-1 security of the grid. Likewise, it could increase the adequacy of the load shedding plan, providing the awareness of the real disconnection of prosumers. Moreover, it could support the TSO during the restoration of the service after a disconnection.

However, the real-time measures at the HV side of the substation, which are currently available for the TSO, are not enough to achieve this objective. Thus, a way to implement the monitoring functionality (the one used in the Italian pilot) is to use aggregation functions to obtain equivalent representations of the grid below each

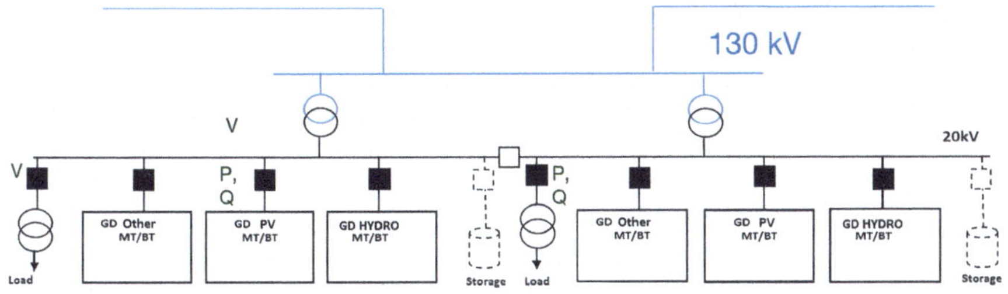

Fig. 6.3 Representation of the MV grid connected at the same primary substation

HV/MV transformer. As represented in Fig. 6.3, the grid fed from the same HV/MV substation is modelled and the generation (represented as GD in the figure), differentiated by type of source and load, is aggregated at that point. This representation of the sources connected to the underlying MV and LV grids enables the TSO to have a better observability of the whole grid to ensure a safe, efficient and reliable management of the system.

Two types of data are necessary to properly represent the electrical grid at the distribution level. On the one hand, the nominal data of installed power is used to describe the structural characteristics of the grid. The installation data must be collected at the primary substation for the different types of energy sources (PV, wind, storage and other sources) and for the load. On the other hand, real-time data are necessary to be aware of the operational conditions of the system. For that purpose, active and reactive power data are sent to the TSO in the form of aggregations, differentiated according to the type of energy source, equivalent to the distributed resources connected at the HV/MV transformer. The aggregations of generation could be composed by measurements in field and estimations of unmonitored plants. In the Italian pilot, almost all the MV production is measured to achieve the required accuracy. The gross amount of load can be measured and/or calculated as the difference between the measurement at the interconnection point and the data collected from generation units.

Real-time data are also used to calculate the virtual capability of the aggregation of MV power plants computed at the TSO-DSO interconnection point, to define the active and reactive availability of the MV resources for ancillary services. The virtual capability is calculated running load flows to evaluate the maximum available capability of the virtual power plant (VPP) in under- and over-excitation. It considers the capabilities of each plant involved in the service, but it also considers the distribution grid constraints. The virtual capability allows the TSO to monitor and control the aggregated area as a unique VPP, while ensuring that the required activation of active and reactive power respects the constraints of the distribution grid.

6.2.3 The Pilot Project

The Italian pilot developed within SmartNet project [2] is a technological experimentation that aims to implement new tools to promote the integration of RES in smart grid systems. The pilot is located in the Ahrntal Valley, in South Tyrol, an alpine Italian region at the border with the Austria, characterized by a wide exploitation of hydropower to produce electricity. This location has several generating modules, of different sizes and connected to several voltage levels. Furthermore, the abundant water flows during the summer, due to the copious winter snowfalls, lead to a high hydroelectric production that often exceeds the local load, causing reverse flow.

The pilot involves the sub-transmission grid at 132 kV and a part of the distribution grid. On the one hand, two hydroelectrical power plants of about 43 MW are directly connected to the TSO 132 kV HV substation. On the other hand, the MV grid connected at the 132/20 kV primary substation includes 23 power plants, with an installed power of 29 MW (27.7 run-of-river hydroelectric, 1.5 biomass and 0.2 PV). The MV grid also includes five local DSOs, characterized by a small number of customers fed by one or more small-sized hydroelectric plants. Due to the behaviour of the subtended grid, the interconnection points with local DSOs are comparable to prosumers with 17 MW of total power consumption.

The consequence of this high hydropower penetration at distribution level is that the active power flows from MV to HV grid. The reverse flow is evident in Fig. 6.4, where the active power at the transformer has negative values for most of the year (passive sign convention) with a peak higher than 30 MW in summer. Figure 6.4 represents the flow in each of the two transformers (red and green) and the total flow in the HV/MV substation.

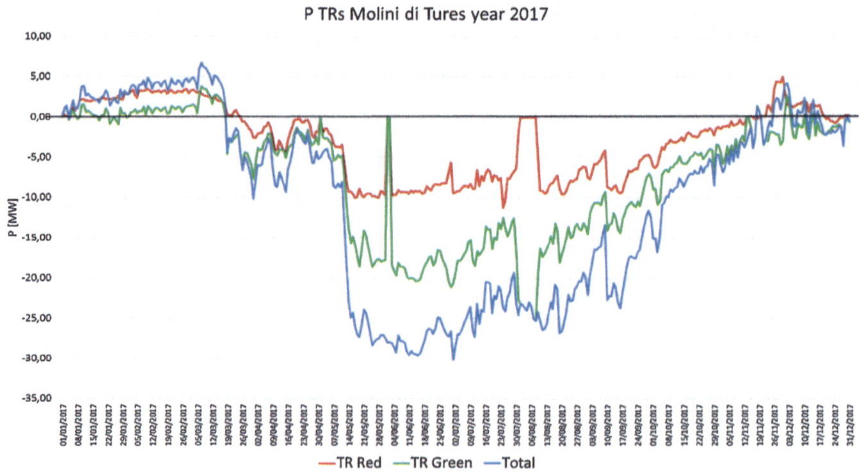

Fig. 6.4 Active power at the HV/MV primary substation during 2017

Within this scenario, the pilot aimed to develop and implement in field two devices to monitor in real time the sources connected to the distribution grid and to test the response of RES power plants in the provision of coordinated voltage regulation and the power/frequency regulation (automatic frequency restoration reserve, aFRR), controlled in a centralized scheme by the TSO. The two devices developed within the pilot are:

1. *High-voltage regulation system* (HVRS), installed in the HV part of the substation, to control the reactive power of the two hydro plants directly connected at the 132 kV sub-transmission grid, which currently do not participate in the hierarchical voltage regulation
2. *Medium-voltage regulation systems* (MVRS), installed in the DSO Operation Centre, to allow the TSO to monitor and control the distributed generation connected to the HV/MV transformers of the primary substation

Figure 6.5 shows the system architecture implemented in field and the data flow among the devices involved in the pilot.

"HS" represents the HV part of the primary substation, "RES" the two hydropower plants connected to 132 kV, "DSO OC" the DSO Operation Centre and "PS" primary substation. HVRS is represented in the upper part of the figure and MVRS in the lower part.

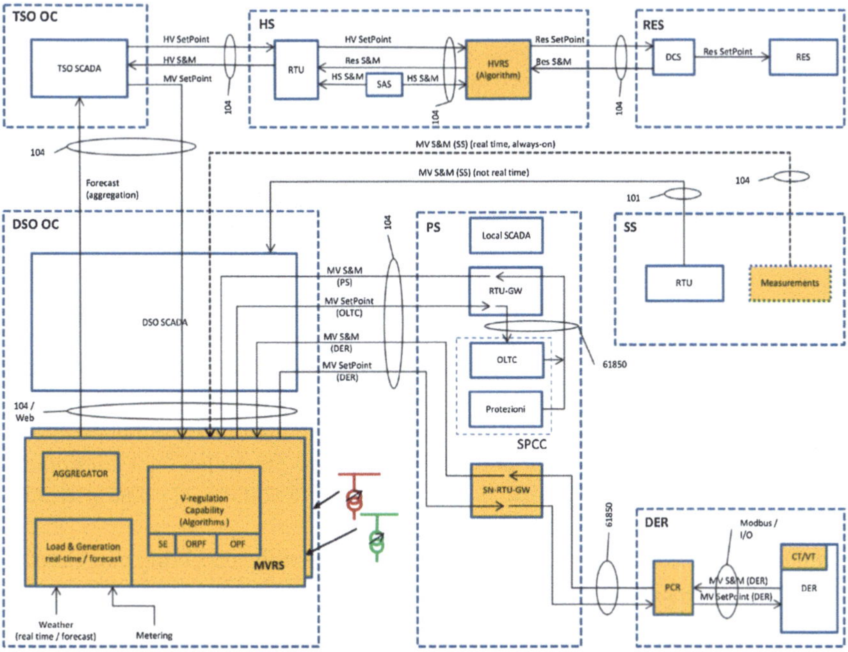

Fig. 6.5 Architecture of the system implemented in the Italian pilot

In order to monitor the 23 power plants and the 5 interconnection points with subtended DSOs, 28 meters have been installed in field. These devices, called Plant Central Regulators (PCR), represent the interface between the power generation module control system and the MVRS. 16 of them are installed at the point of connection of the power plant with the grid, to measure the active and reactive power exchange, and 7 of them are installed at the terminal of the generators, to control 7 of the biggest hydroelectric plants (of about 22 MW total).

The HVRS and the MVRS, developed by the technological partners in the Italian pilot, were created to test the following functionalities:

- The coordinated voltage regulation provided by the hydropower plants connected at the sub-transmission grid
- The computation of the dynamic capability of the aggregation of power plants connected at the distribution grid
- The voltage regulation provided by the aggregation of distributed generators involved in the pilot
- The power/frequency regulation provided by the aggregated distributed generation

The first functionality was implemented by the HVRS, whereas the rest were realized by the MVRS.

6.2.4 Description of HVRS Functionalities

The HVRS device was developed to control, in a coordinated way, the reactive power absorbed/injected by the four generators, in order to regulate the voltage at the busbar of the 132-kV substation, following a set point sent from the TSO's control room.

The voltage regulation is a hierarchical service in Italy, where only the pre-qualified power plants can provide secondary voltage regulation, i.e., tele-controlled regional voltage regulation. Only big-sized, programmable power plants connected at the transmission grid and equipped with specific devices designed to provide the reactive power service can pre-qualify.

The HVRS was implemented in order to allow hydropower plants connected at the sub-transmission grid to provide this service. For that purpose, the device computes and sends to the TSO the reactive power availability of the system composed by the four generators in over- and under-excitation. Then, the TSO controls the power plants by sending a reactive power set point or a voltage set point referred to the HV busbar of the substation.

In the first case, the set point is a percentage value of the capability calculated in current operating conditions, to be interpreted as follow:

- Value between (0, +100%] indicates the condition of over-excitation, i.e. represents the reactive power that must be provided by the plants in order to increase the voltage.
- Value between [−100%, 0) indicates the condition of under-excitation, i.e. represents the reactive power that must be absorbed by the plants in order to reduce the voltage.

The algorithm implemented in the HVRS shares the desired reactive power (Q) level among the four synchronous generators and sends the calculated values to the distributed control system (DCS) of each power plant.

In the second case, the set point is the optimal voltage value expressed in kV. In this case, the HVRS converts the set point in a reactive power command on the basis of the voltage error, defined as the difference between the voltage set point and voltage measurement. The correlation between the production/consumption of reactive power and the voltage error defined by the technological partner is a cubic law (Fig. 6.6). This cubic law can be parameterized to have an adjustable control to change the gradient of the trends. This way, the correlation between the two variables can be modified: for the same optimal voltage, if the gradient is increased, the reactive power contribution of the generators also increases.

Figure 6.7 shows the result of a test carried out to check the behaviour of the HVRS. The orange line is the voltage set point sent by TSO, i.e. the optimal value of the voltage at the busbar, while the blue line is the voltage trend at the busbar.

Although the response by power plants had some issues (delays and overshoots) and a limited benefit on the voltage busbar, the HVRS allowed the TSO to coordi-

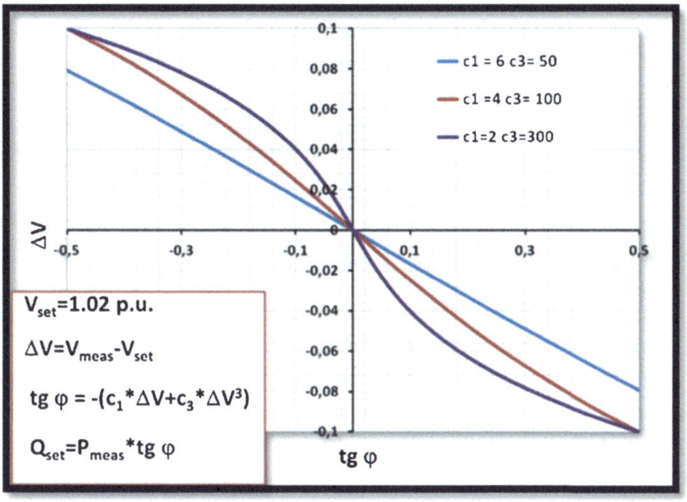

Fig. 6.6 Cubic correlations of the control law that express reactive power contribution as a function of voltage error

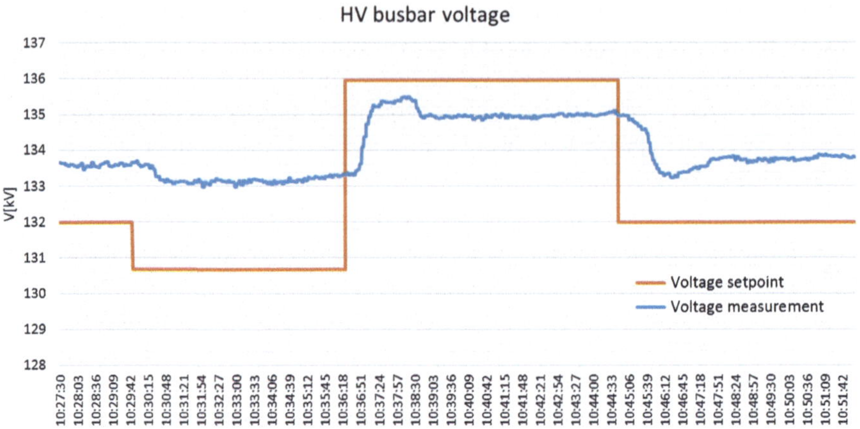

Fig. 6.7 Effect of the reactive power regulation on the HV busbar voltage

nate the reactive power exchange of different power plants to satisfy the requirements in a given area. This way, reactive power loops between the groups were avoided and, thus, the system prevented wasting reactive power resources.

6.2.5 Description of MVRS Functionalities

The first step of the MVRS algorithm is the computation of the dynamic capability of the aggregated distributed generation subjugated to the device, which is considered as a VPP. This capability is defined by the active and reactive power limits of the VPP, and, based on those limits, the MVRS provides the TSO with information on the availability of active and reactive power that can be activated to provide ancillary services. These limits are defined by running continuous load flows to evaluate the maximum available capability of the VPP, considering the nominal capability of each power plant, the operational status and the constraints of the grid.

To ensure the priority of distribution grid constraints, the VPP is declared available for the provision of ancillary services only in absence of distribution grid violations. If the device detects a violation, the MVRS tries to solve the violation acting on the PCRs in power plants and on the transformer's tap changers. This allows real-time control of distributed generation, while respecting grid constraints.

Once the dynamic capability is computed, the MVRS also allows distributed generation to participate in both voltage and power/frequency regulation.

Regarding voltage regulation, the MVRS implements the same approach as the HVRS and tries to regulate the HV side of the MV/HV transformer at the primary substation, by controlling the reactive power exchanges of the seven generators involved in the pilot and connected to MV.

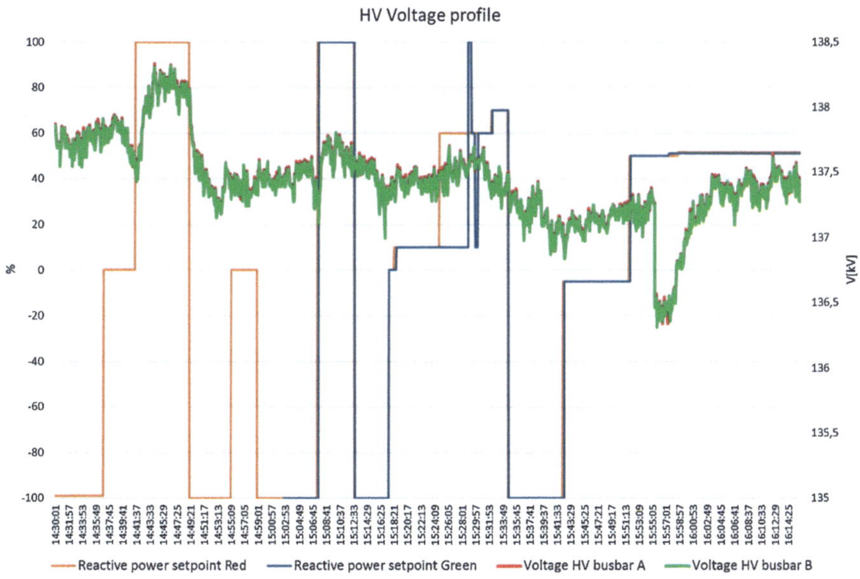

Fig. 6.8 Trend of the voltage at the HV side of the primary substation during tests

Figure 6.8 shows the results of a test carried out to compare the reactive power set point sent by the TSO to the MVRS and the voltage trend. As in the HVRS case, the voltage reacted to the set point modification, despite the small delays in the response due to the length of the data-flow chain and the operation of other elements of the grid that affects the HV busbar (as happened at 15:55), thus reducing the benefit in the transmission grid management. Nevertheless, the coordination of the reactive power exchange of distributed generation can reduce the waste of reactive power reserve, with particular interest for future scenarios when the phase out of coal-fired power plants will reduce the resources available for voltage regulation.

The benefits of the MVRS are more evident for the DSO to control the voltage rise effect along the feeders of the distribution grid. As shown in Fig. 6.9, the MVRS contributes to maintaining the voltage profiles of the distribution grid within standard operational limits, even when it is not operated by the TSO.

The last functionality of the MVRS is the power/frequency regulation. The current Italian regulation has recently opened the replacement reserves (RR) market to the DER (del. 300/2017/R/eel), while the aFRR is still provided only by programmable power plants connected at the HV grid and of a size greater than or equal to 10 MW.

The aFRR service is provided by modifying the active power production around the programmed value on the basis of a tele-signal received by the national regulator and the band made available for the regulation by the power plant. The set point is calculated automatically considering the deviation from the nominal frequency and is expressed as a percentage between 0 and 100 where:

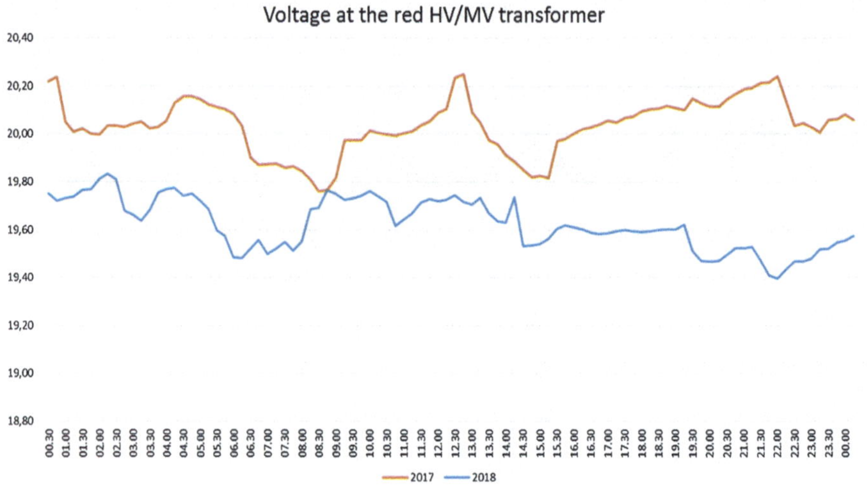

Fig. 6.9 Comparison of voltage trend at the transformer with (blue) and without (orange) the MVRS

- 0 represents the lower limit of the band and, then, the minimum production power made available for downward service (to reduce the generation).
- 50 represents the midpoint of the band, and it coincides with the programmed production (no activation).
- 100 is the upper limit of the band, and it represents the maximum production power made available for upward service (to increase the generation).

The Italian pilot aimed to test this service involving an aggregation of non-programmable power plants connected at the distribution level. MVRS calculates and sends to the TSO the programmed production and the band available for the regulation, by considering the value declared by each power plant included in the VPP. The tele-signal is sent to the MVRS every 4 seconds, which splits the command among the regulating power plants.

In this application, two simplifying aspects were considered:

- In absence of a baseline defined by the market, the last active power measurement before the start of the test was considered to be the programmed value.
- The power plants provided only downward reserve, to avoid maintaining a margin between the operating point and the maximum production.

The tests were performed by sending a level signal with a ramp profile, composed by a ramp-down to reach the minimum production made available and a ramp-up to return to the initial programme value as reported in Fig. 6.10.

The tests provided promising results, with the activation of seven power plants and a variation of the production of more than 6 MW. Regarding the quality of the regulation, the dynamic response did not comply with the technical requirements of the service, due to delays in the communication and the inaccurate regulation of the

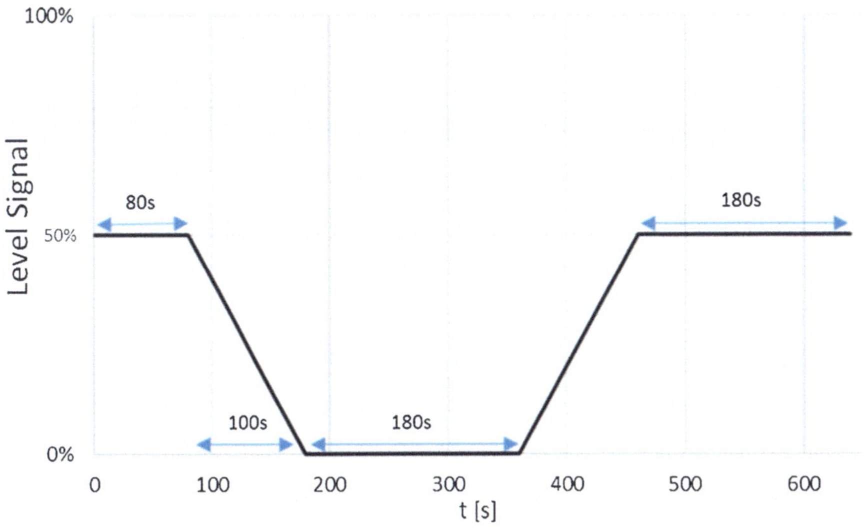

Fig. 6.10 Ramp of the level signal used for the power/frequency regulation tests

power plant governor. Moreover, the tests showed that the reliability and the quality of the regulation of the virtual power plant at the interconnection point do not solely depend on the single power plant performance, but the trend is influenced by other elements of the grid, which are uncontrolled and unforeseeable.

An example is reported in Fig. 6.11, where the blue line is the expected contribution calculated from the percentage set point sent by the TSO and the orange line is the real contribution of distributed generation, which is calculated by subtracting an offset value to better appreciate the trend. In the lower part of the graph, the trend of the dynamic error has been reported in comparison with the limit value used in the acceptance tests of the aFRR service (10%). The error increases with increasing response inaccuracy and delay.

6.2.6 Lessons Learnt

The implementation and the testing of the pilot reached very important technical goals, although some crucial limitations were also identified.

Regarding the system architecture, the central role of the communication chain was highlighted. The data flow involves the TSO's control centre, the Substation Automation System (SAS), the HVRS and the control systems of power plants. Although the same communication protocol was adopted, the need of dedicated assessments between devices of different manufacturers to obtain a reliable communication among devices was evident. The availability and reliability of the telecommunication network is also needed, especially when the controlled power plants

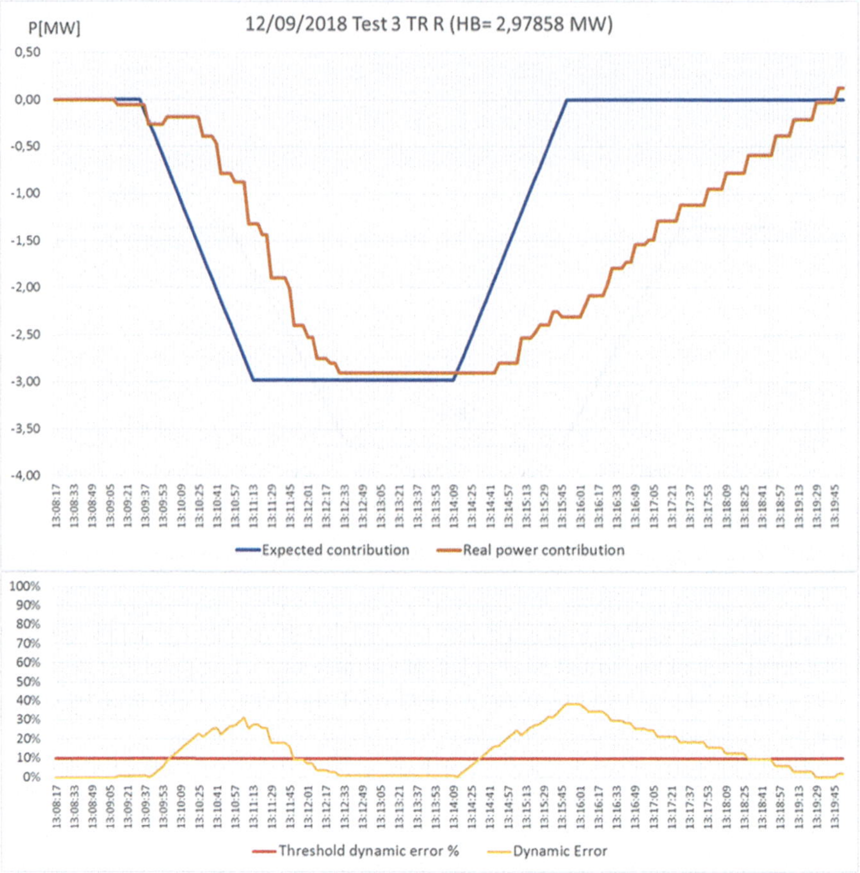

Fig. 6.11 Example of trend and analysis of the HV contribution of the virtual power plant connected at the transformer

are located in remote places. Furthermore, it is important that the communication respects TSO's safety and quality standards.

From a technical point of view, the operation of the HVRS showed the feasibility to coordinate the reactive contribution of different plants connected at the same substation, even if they are based on different technologies (RES and traditional generators), when contributing to the management of the grid operated by TSO. The provision of the voltage regulation strongly depends on the dynamic response of the power plants, which needs further assessment to update the configuration of the local control equipment to enhance the response and reduce overshoots and delays. Likewise, the impact on the transmission grid voltage is not comparable with the contribution currently provided by traditional power plants, due to the technical limits of the RES power plants and the resistance of the grid against activation in the sub-transmission grid.

Regarding the functionalities of the MVRS implemented and tested, the challenges encountered with the HVRS were amplified. The pilot project showed technical barriers for the integration of distributed generation in the grid management.

Firstly, it was very clear the necessity to have continuous and accurate monitoring and transparent access to the MV and LV levels of the grid in order to ensure a safe and reliable management of distributed resources, particularly to handle non-programmable sources. Furthermore, the 20 seconds update of the aggregated measurements at the interconnection point, which is required for the observability functionality, is not consistent with the monitoring needs of the provision of ancillary services. In fact, the set point is sent every 4 seconds, and, with a 20-second sampling rate, the response of the aggregation cannot be verified, so it is not compatible with safety assessments.

The pilot showed the technical feasibility of controlling the active and reactive power of distributed generation, through a centralized scheme. The test results highlighted the necessity to improve the technical performance and the capabilities of RES power plants located at distribution level, because the response is not complaint with the requirements of the services. Furthermore, the non-programmability of the resources and the need to find a way to overcome the unpredictability of RES pose an additional constraint: it is not possible to guarantee the contribution of RES to the provision of ancillary services without a production plan or without considering the combination with other types of flexibility, which can compensate the performance errors presented by the tests. Such results discourage the replacement of traditional plants for the provision of ancillary services, for the safe operation of the system.

An important result obtained is the possibility of managing the distributed generation and activating ancillary services in the distribution grid without incurring in violations in the operational limits of the distribution grid. The computation of the real-time capability takes into account the distribution grid constraints, and, in case a violation is detected, the VPP is declared unavailable for the provision of the services for the TSO.

Finally, the communication chain shows several critical components, because the interposition of numerous devices results in delays in the data flow. Moreover, the signal conversion of each device causes approximations that lead to inaccurate response and data transmission.

6.2.7 Conclusions

In conclusion, the pilot provided an interesting preliminary study to evaluate the technical feasibility of the proposed systems to monitor and control RES and distributed generation. Moreover, it also opened the door to future improvements to understand how the TSO can exploit new flexibilities to support the management of the power system.

Regarding voltage regulation, the experimentation showed that the effect of big-sized plants connected in the area prevails over the contributions of small-sized hydropower plants, especially if they are connected at the distribution level. Although marginal benefits for the voltage at the transmission grid were found, the main advantage is the possibility to reduce the recirculation of reactive power among generators by coordinating the reactive power exchange of the generators, in order to avoid the waste of regulating resources.

A successful application of the MVRS operated by DSO is the local voltage regulation, which led to improvements in the voltage profiles along MV feeders, through reactive power regulation of distributed generation. The pilot proved the feasibility of coordinating the reactive power exchange between the TSO and the DSO through coherent reactive power activations. In any case, further analyses are needed to improve the dynamic response so that these small-scale power plants comply with current requirement for the provision of the service.

Regarding the power/frequency regulation, the tests showed good results on the amount of active power activation, because they allowed for the successful activation of 6 MW in the distribution grid. The experimentation also verified the possibility for the TSO to activate services from distributed generation in accordance with grid constraints. In any case, the dynamic response of each generator and, especially, of the entire VPP did not comply with the requirements of the aFRR service in terms of delay and accuracy.

The main value of the pilot was getting the chance to highlight future opportunities, on the one hand, for manufacturers to improve the performance of the regulation and of the communication chain (for instance, regarding ICT and power plant controller) and, on the other, for system operators to deepen benefits and challenges in the management of renewable energy sources.

6.3 Aggregator Focus: Indirect Control for Demand Aggregation

6.3.1 Challenges in Unlocking the Flexibility

The use of renewable, volatile energy sources like wind and solar energy is necessary to reach the goals from the Paris 2015 agreement. Denmark has the world's highest share of variable renewables, such as wind and solar, in the power system. In 2018, wind contributed to 44% of the total electricity consumption and wind generation exceeded demand in 5% of hours of the year, based on studies conducted by the Danish TSO [3].

The national target for Denmark is to have at least 50% of the electricity consumption from wind and to eliminate fossil fuel plants by 2050. In order to meet those targets, Denmark can take advantage of the six electricity interconnectors to neighbouring power systems, which are an important source of flexibility. However, an efficient transition to a low-carbon society based on such intermittent RES calls for a change to an energy system where the demand follows the production within reasonable-sized temporal and spatial resolutions. This requires a development of new methods to unlock flexibility at all levels of the energy system of the modern society.

At the same time, the energy system is evolving from a centralized power system into a complex set of integrated or coupled energy systems (electricity, thermal, gas) at scales that include customers, cities and regions. By integrating energy systems, advantage can be taken of the synergies between different energy carriers, such as electricity, gas and heat, to secure a safe and resilient operation of the grids in real time. This leads to a complex system where the computational efficiency must be tailored to the need for a near real-time matching of energy demand with production.

Although existing ancillary service (AS) markets and mechanism have successfully served power systems in the past, it lacks certain features and requirements to cope with the emerging requirements. Besides being simple and secure, the existing AS markets over simplify assets' operation to linear price-quantity block of bids. They simply ignore the inherent dynamics and uncertainty related to the flexibility of the underlying systems and equipment. Moreover, the conventual framework is understandably slow, because a large-scale optimization problem with thousands of variables and constraints along with power flow should be solved in every relevant time interval. Moreover, existing AS markets are designed to only procure services from conventional power plants. It is not possible – or very difficult – to extend the market to end users' flexibility resources in this format because it will require managing bids and activation of millions of new DER which is not practical. Furthermore, since the market is only designed for electricity resources, it is technically impossible to directly incorporate flexibility, which arrives from the interplay with other energy carriers of the future integrated energy system.

Conventional market principles characterize flexibility by the elasticity of demand, and the methods considered for balancing the systems and for providing AS are all two-way communication systems, like the transactive energy and peer-to-peer approaches.

An important technological advancement incorporated into Danish pilot [4] is the field test and proof of concept of DERs using unidirectional communication as well as forecasting and control-based technologies. The concept called the Smart-Energy Operating-System (SE-OS) will be outlined in the next section. It will be argued that this new advancement leads to a possibility for very fast calculations and, most importantly, it leads to a possibility of using new principles for modelling the flexibility which reflects the need for a description of dynamics and stochasticity and which gives us a possibility for using rather simple control-based techniques as an alternative to the computational demanding large-scale optimization.

For that purpose, the system will need to use the real-time data generated from various sources, including the meteorological, electricity market and consumer's demand behaviour data, when integrated with existing solutions and algorithms. The exploitation of the data requires a new physical and technological setup that includes installing Internet-of-things (IoT) devices, providing communication links and using machine learning, artificial intelligence and new market approaches to facilitate the provision of flexibility at large scale.

6.3.2 *Smart-Energy Operation-System (SE-OS)*

In such unidirectional context, two different control mechanisms, i.e. direct and indirect control, can be implemented. The former is a direct control signal requesting the DER to turn on/off based on the optimization done at the aggregator's side, while the latter, which is adapted to Danish pilot, allows DER to perform economic optimizations and leaves to the DER themselves the ultimate decision to get activated or not.

The Smart-Energy Operating-System is a framework for implementing flexible energy solutions [5]. It consists of both direct and indirect (mostly price-based) control of the electricity load. It contains methods to implement solutions for handling ancillary service problems. Most importantly for the SmartNet project, the concept is equipped with, for instance, a methodology for price-based control of electricity load in future electric and integrated energy systems.

The SE-OS, as shown in Fig. 6.12, includes stochastic optimization layers based on models representing aggregated consumption on various spatial and temporal scales [6]. On large scales (e.g. day ahead and for a large region), which is represented in the upper part of Fig. 6.12, the optimization is based on conventional market bidding principles, while on more fine resolutions (e.g. for the next minutes within a DSO network), which is presented in the lower part of Fig. 6.12, the optimization is implemented as a model-based control. In this way, the dynamics and stochasticity of the flexibility can be taken into consideration, but it calls for new approaches for describing the flexibility.

For example, SE-OS has been used to implement flexible and smart grid enabled solutions for wastewater treatment plants, interactions between power and heat, smart energy control of supermarket cooling systems, new solutions for control of heat pumps and methodologies for using thermal mass in buildings and district heating systems as an energy storage [7]. Price-based control [8, 9] is an important part of the SE-OS framework. Danish pilot has benefited from SE-OS by implementing the top-down, one-way communication from aggregators to DER using price-based control method.

The control or penalty signal is usually a price. However, as demonstrated in the Danish pilot, other penalty signals, such as the real-time CO_2 content of the electricity, can also be used. In this case, the SE-OS controller leads to a CO_2-efficient controller, since the total CO_2 emission linked to the electricity consumption will be minimized. Depending on the selected penalty function, the controller can provide cost, emission or energy efficiency, as illustrated in Fig. 6.13.

As shown in Fig. 6.13, multilevel controls of the SE-OS can be implemented. In this case, the upper levels are used to generate prices related to various types of

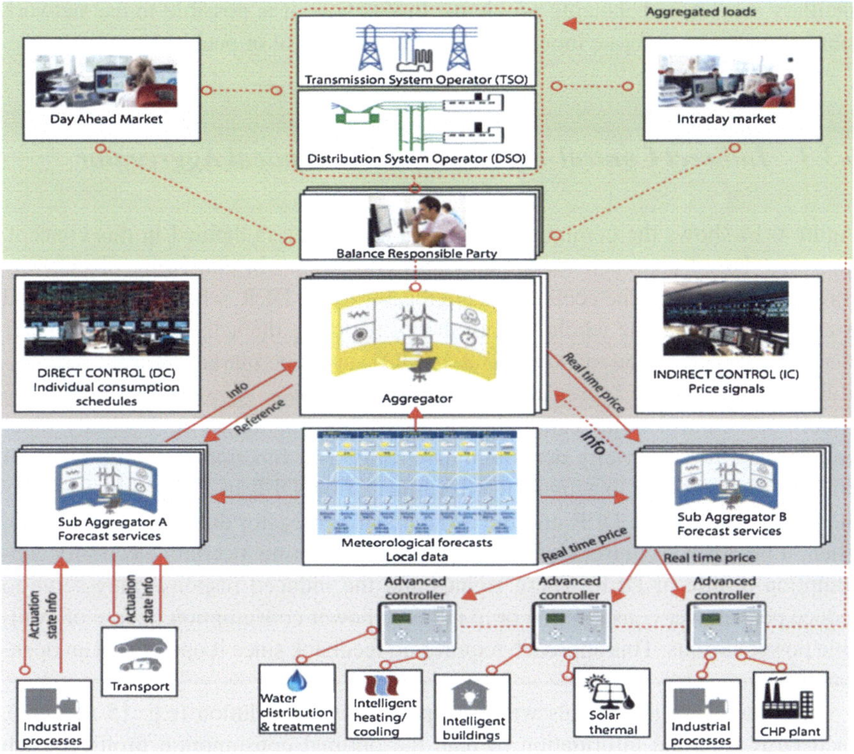

Fig. 6.12 The SE-OS framework

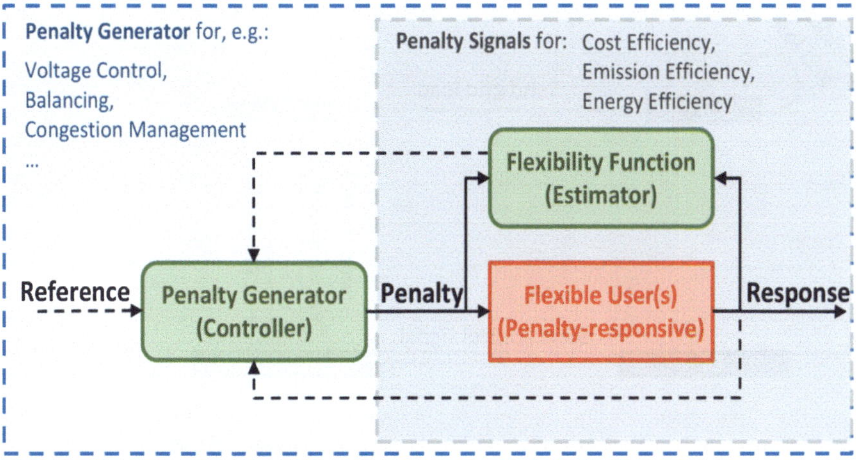

Fig. 6.13 The penalty and flexibility function part of SE-OS

ancillary service or balancing problems. In this way, it is possible to use network measurements to generate inputs for, e.g. voltage control or peak shaving [9].

6.3.3 Indirect Control Through the Economical Aggregator

Figure 6.14 shows the communications among the actors defined in this concept. The price-based approach implements an indirect control consisting of one-way communication from the economical aggregator to the DER, where the price signal is used to influence the whole load of the DER during the activation period. After clearing the market, the market operator (MO) sends the market clearing information to the economical aggregator. In turn, the economical aggregator calculates the price-based control signal estimating the flexibility function. The flexibility function predicts the electricity demand dynamically as a function of a time series of prices. The purpose is to activate the flexibility of the DER in a way, which creates the most value for the DER and the economical aggregator during the next hours. Then, it broadcasts control signals to the DER, prompting a certain electricity consumption profile of DER. These signals and the induced response may serve to reduce peak power consumption or to increase power consumption in case of available power surplus. This approach requires no feedback since it operates in an open-loop scheme.

After receiving the signals within a specific time resolution (e.g. 15 minutes), each DER uses the information to plan the optimal consumption profile, which results in the lowest electricity bill, while staying within the temperature boundary conditions. Before reaching the next time step, the price for the next time step is sent

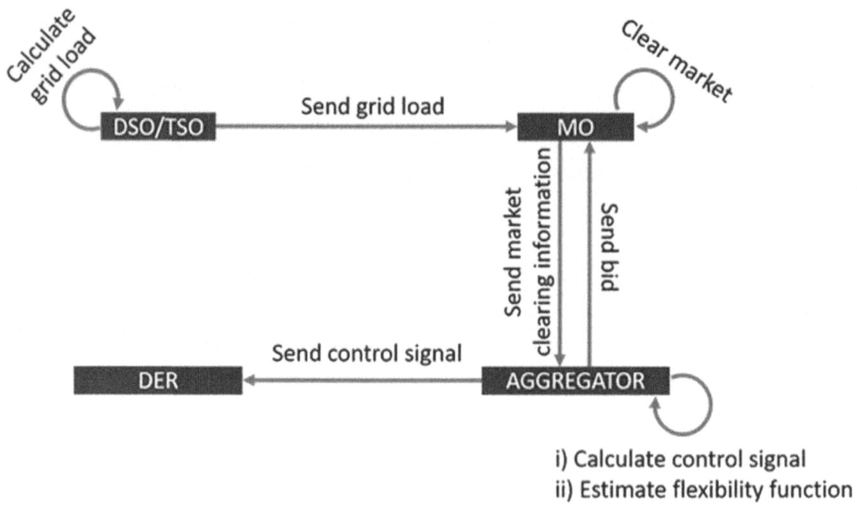

Fig. 6.14 Main communications for indirect control of DER through the economical aggregator

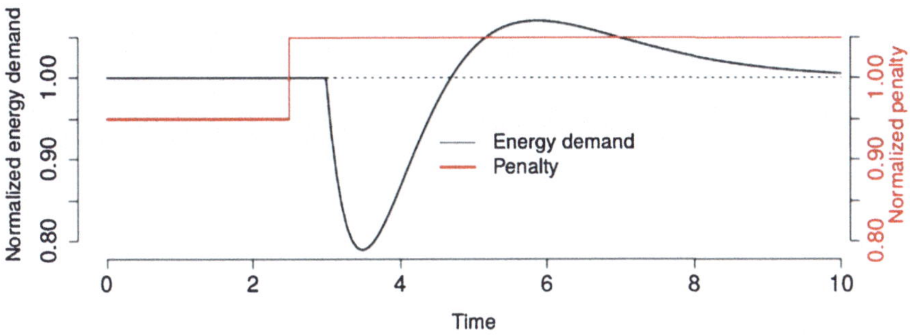

Fig. 6.15 Example of a flexibility function

from the economical aggregator, including an updated price forecast. Each DER updates its consumption profile for each time step. This results in a quite simple unidirectional communication system, which does not require the commitment of the DER. It lets the DER optimize their consumption continuously. One challenge is, however, for the economical aggregator to predict the response from the DER at a given price signal.

By establishing a price generation mechanism, the economical aggregator determines the optimal real-time price signal based on the estimations of the aggregated response, the so-called flexibility function [10]. Such estimations are based on historical data, and the characteristics of the response can be tailored to specific needs. In International Energy Agency (IEA) Annex 67 "Energy Flexible Buildings" [6], for characterizing the aggregated flexibility response on a step change in price, the authors use a step response function as illustrated in Fig. 6.15.

6.3.4 Grey-Box Modelling

In this section, the grey-box modelling principles used for identifying proper models for the heat dynamics of the summer houses used in the Danish pilot are briefly described. The principles are generic and have been used for estimating similar models for ordinary buildings, greenhouses, wastewater treatment plants, district heating systems, supermarkets, etc. For the Danish pilot, the grey-box models are used for both short-term simulations of the temperatures of the pool and for model predictive control.

Conventional, physical *white-box models* based on first principles are most often not useful for operational purposes, since the number of equations and parameters implies that the computational time does not fit with the real-time needs, and furthermore the number of parameters is often too large to facilitate an online learning from data. Other models identified using machine learning or statistics are most often not linked to the physics of the system, and hence such models are called *black-box models*.

Therefore, simplified, so-called grey-box or cyber-physical models are more useful for online applications. Such models are often formulated in state-space form as discretely observed stochastic differential equations [11, 12]. This formulation provides several benefits for online operations like forecasting and control:

- The model states can be estimated and simulated in almost real time.
- Model assimilation is possible, meaning that the parameters can be estimated based on real-time data.
- Uncertainty of the evolution of the states can be specified, which implies that risks measures can be taken into consideration.

Consequently, the best choice for many smart energy applications is to consider grey-box models. The class of *grey-box models* bridges the gap between physical and statistical modelling.

Another main advantage of grey-box models is their ability to couple detailed physical models to data and thereby providing an insight into the detailed physics and dynamics of a system. A grey-box model is formulated as a continuous time model for the states of the system, together with a discrete set of equations describing how the measurements are linked to the states. This is often called a continuous-discrete time state space model (see Chap. 10 in [13] for further details about state space models). The continuous time formulation of the dynamics ensures that prior physical known relations, which typically are given as differential equations, can be used as a part of the model formulation.

For describing the thermal inertia of buildings, most often the so-called RC network models are considered. These models belong to the class of linear grey-box models, which is the classical dynamical model most frequently used for buildings and building components. However, modern buildings (e.g. buildings with a lot of glass or natural ventilation) and advanced walls (e.g. walls with PV-integrated panels) contain non-linear phenomena like those related to radiative heat transfer, free convection, etc. For such more complicated phenomena, the class of non-linear grey-box models must be considered.

For the summer houses with a swimming pool, the grey-box model shown in Fig. 6.16 was used, which is further described in [14, 15].

The parameters of the model are estimated adaptively, and in this way the changing dynamical behaviour of the system is accounted for. One example is the amount of water in the pool, which then is indirectly estimated in real-time.

6.3.5 Implementation Setup

The Danish pilot aimed at exploiting the flexibility in indoor swimming pools to provide services for the operation of the grid, based on the SE-OS framework. To do such studies, it is essential to break down the implementation process into numerous technical steps. These steps can be categorized into technological requirements,

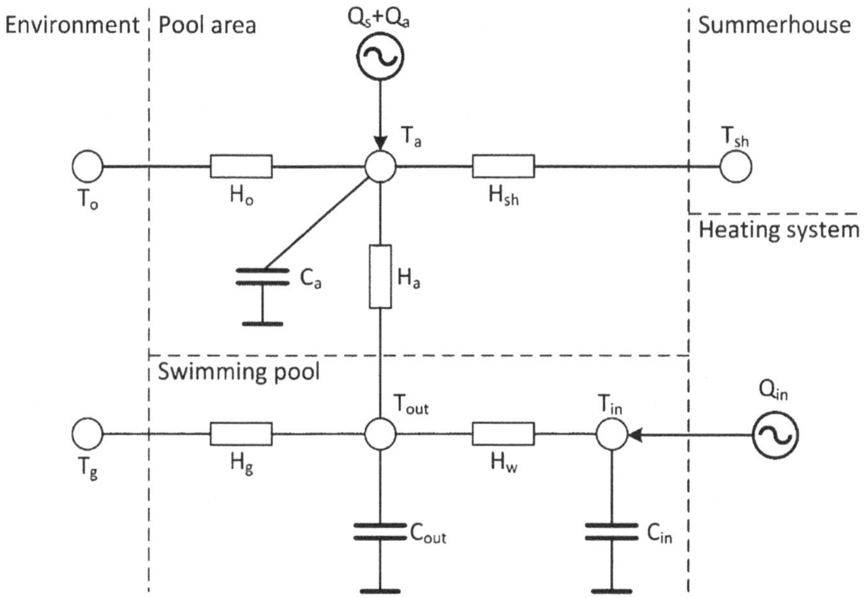

Fig. 6.16 Grey-box model for the summer houses with a swimming pool

availability of meteorological data, having access to the electricity market price data (market operator), real-time data from flexibility providers, visualization tools and having a reliable communication network. A similar requirement analysis was adopted by the Danish pilot to ensure that the available tools, technologies and open-sources software facilitate smooth and seamless interactions between the servers and physical devices.

Technological requirements refer to choosing the right hardware and protocols to handle the exchange of information from the IoT devices to the central control software systems that send the control signal to the underlying systems and equipment. A dedicated gateway system was designed and developed (called SN10) for the Danish pilot to act as a connector between the cloud and the installed sensors and controllers in the swimming pools. The pilot deployed a Representational State Transfer (REST) API with a Json format for transferring the data back and forth from the SN10 to the server.

The availability of *meteorological data* is critical for predicting and calculating the set points by the control algorithms. The meteorological data include historical weather temperatures, the forecast data for the next 24 hours and indoor temperature recordings by the sensors in the buildings. In the Danish pilot, these data were provided by a third party to the control algorithm developed to provide the signals to the controllers in the summer houses. Periodically, the control algorithm requested the weather data and the SN10 readings in the swimming pool to calculate the next time-horizon control signals.

In addition to meteorological data, it is also very important to have access to the *electricity market price* data. Using the actual price data, combined with the meteorological data, the flexibility function provided the optimal price signals.

Cloud services are the backbone of such setups as they provide centralization, enabling scalability and ensuring reliability of services. In the pilot, Amazon Web Services (AWS) were used to host the market operator software, a dedicated cloud service for the main control software that depend on the market data, meteorological data, the summer house occupancy data and the temperature data of the swimming pools. The data exchange requires a stable *communication network* in the summer houses. Ideally, wired connection would work best, since its reliability is very high, or, alternatively, using the building WiFi. However, in the Danish pilot, cellular network was used to transmit and receive data from the servers. Some of the limitations of a single solution include lack of telecommunication signals and summer houses located in the remote areas with limited GSM signal, which resulted in some disruption of automation services in the swimming pools. Based on the learnings of the pilot, it was evident that to overcome the communication issues, the most optimal solution is to have a dedicated broadband connection for the SN10 and the IoT devices in the summer houses.

The *controlling approach* was based on designing two separate algorithms, a local controller (LC) and a model predictive controller (MPC), which generated the optimal temperature set points (T_{sp}) to be sent to the LC. Using the measurements obtained from the summer houses, which include booking status, as well as electricity and weather forecast, the controllers aim to minimize the operational cost. Figure 6.17 shows the overall design concept of the control system developed and implemented for the summer houses in an abstract level.

6.3.6 *Results and Impact*

Upon completion of the Danish pilot, it was evident that by using the tools and technologies available combined with newly developed algorithms, energy flexibility could be harvested from the existing resources. In fact, the Danish pilot demonstrated that deploying such methodologies and setup reduces the CO_2 emission by at least 10%. Similarly, the summer house owners can reduce their annual utility bill by 8–12% by using the setup deployed in the pilot. These savings can be even higher

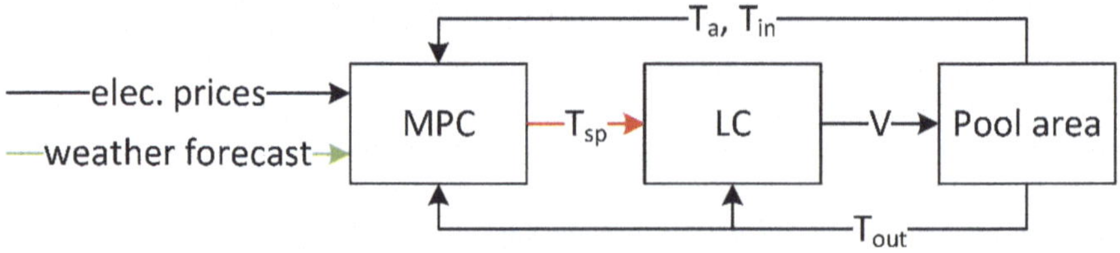

Fig. 6.17 Controlling approach

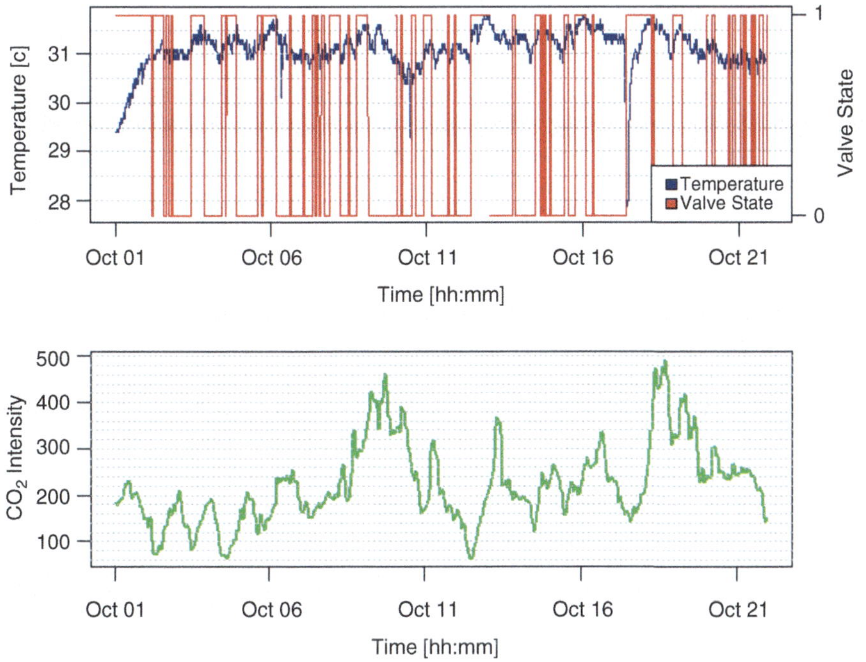

Fig. 6.18 Example of CO_2-based control [4]

if an improved setup is implemented, based on the lessons learned during the course of the pilot execution.

Figure 6.18 shows an example of application of CO_2-based control. In the upper part, the temperature is shown in the blue and the heating state in red (1 if on, 0 if off) for a period of 21 days. The lower part shows the CO_2 intensity recorded for the same period. By looking at the figure, it is evident that the heating comes on when the CO_2 intensity is low. By comparing the CO_2 intensity while the heater is on to the ordinary operation, the total CO_2 reduction can be calculated which, in this case, was 9.6%.

In addition to the technological developments, policy and regulation also play a significant role in the success of pilots, like the ones deployed in SmartNet. For instance, consumers may be encouraged to participate in these pilots by providing them tax break or giving them lower electricity tariff. A solution for the exploitation of flexibility like the one presented in the Danish pilot has limited reachability if regulators do not support it, since, for some smaller consumers, the savings are not significant enough to encourage them to participate in such schemes. In addition, for the larger consumers, such as supermarkets or manufacturing sector, the incentives complement the savings made from deploying the flexibility setup. Denmark is presently providing extra funding to the DSOs and research institutes to further explore the concepts and methodologies adapted in the Danish pilot and to expand it to wastewater treatment plants and other sectors of society.

6.3.7 Conclusion

The Danish pilot demonstrated that it is possible to gain extra flexibility from existing resources by utilizing the information and real-time data available from the assets, combined with having access to electricity market and meteorological data into novel model predictive control algorithms. However, for such solution to work, there is a need for a dedicated ICT solution with stable communication links that includes cloud services, forecasting models, future occupancy plans and, most importantly, the incentives for consumers to participate in such solutions. It is important to have a dedicated ICT infrastructure designed and developed for the success of the solutions because it allows gaining greater reliability and facilitates the scalability in the future.

In general, the solution implemented in the Danish pilot was pioneer in using the price signal concept in a real-world scenario setup and, by doing so, in leveraging energy flexibility. Applying the grey-box modelling techniques combined with the flexibility function developed for the pilot proved to be extremely important to understand the dynamics of the energy usage in the summer houses. In this kind of solutions, where the aim is to provide flexibility from the demand side, the activation of such flexibility could also help reduce both the procurement of conventional reserves and the additional investment needed. In general, when considering energy flexibility, it is important to consider the stochasticity of it, as it does not follow the traditional power grid production and demand concept, since flexibility is dynamic. This dynamic behaviour could be a problem for an aggregator willing to offer flexibility to participate in the current electricity market and would put them in a weak position versus conventional aggregators. Therefore, the electricity market may need to be restructured to better integrate smaller aggregators who only provide flexibility.

6.4 DSO Focus: Creation of Local Markets for Improving Distribution Grid Operation

6.4.1 Challenges and Concepts

The share of highly variable production from RES is also increasing the complexity activities in the Iberian market at TSO level. On the one hand, the Iberian peninsula is almost an electrical island in the European power system, as recognized by ENTSO-E for all network development scenarios in 2030 [17] and presented in Fig. 6.19. As a result, the Spanish system cannot benefit from exchanging power with neighbouring countries, as, e.g. Denmark does.

On the other hand, balancing the system is an increasingly complex (and costly) activity, due to the importance of the highly variable RES production. There are relatively frequent episodes of high variability in the production and share wind

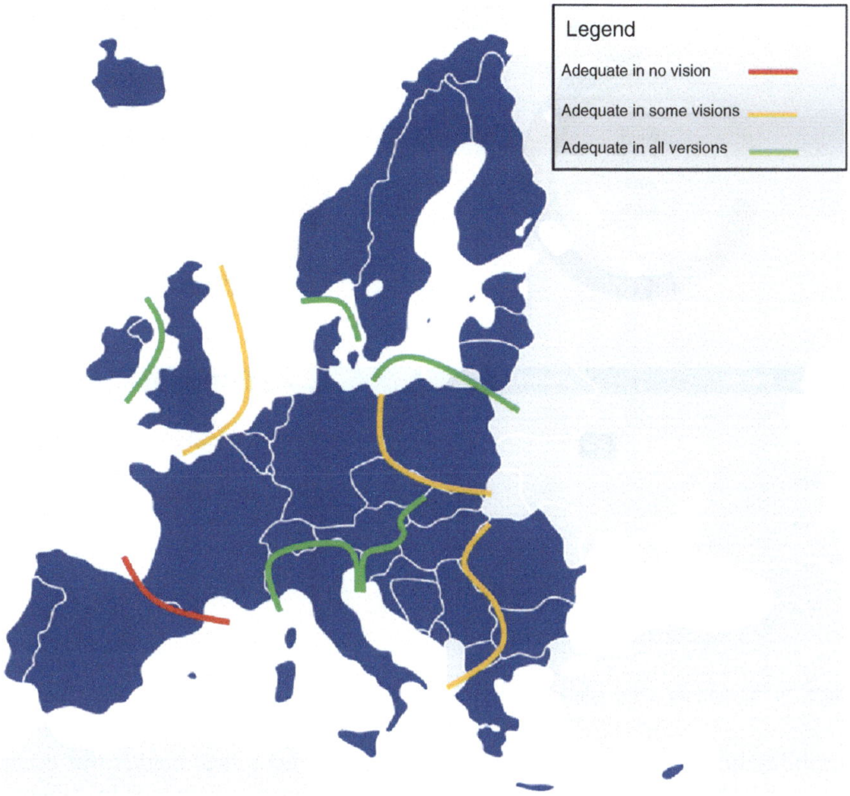

Fig. 6.19 2030 transmission adequacy [17]

power in the Spanish generation mix, like the one presented in Fig. 6.20, where wind power increased its production from roughly 2 GW at 14:40 (contributing to 6.7% of demand) on 26/01/2019 to almost 11 GW (43.5% of demand), at 02:10 on 27/01/2019, less than 12 hours later.

Balancing and associated AS become important methods to solve this uncertainty of generation. In addition, the increasing number of DER and consumers may lead to grid congestion issues, even at distribution grid level.

The grid digitalization is another challenge for the distribution systems; the new paradigm forces the DSO to provide observability and controllability to the grid. A high level of digitalization can provide early fault detection to the DSO and the possibility to take actions to solve any issue, but, at the same time, the DSO should be able to manage and organize all network data and information.

In line of this new paradigm, the Spanish pilot was defined and deployed [16]. The aim was to test the capability of small-scale DER to provide AS which could help improve the operation of the distribution grid. By exploiting this flexibility, the DSO can obtain profits from its traditional business model, grid operation, but the aggregator and the DER owner do not, so they must be compensated for the

Fig. 6.20 Variability of the generation mix in the Spanish power system [18]

flexibility they provided to the DSO. For that purpose, a local market was established, which established the compensation for the services provided by different DER in a competitive manner.

Since the Spanish TSO was not part of the SmartNet project, the "shared balancing responsibility" market model was chosen. Under this model, the TSO and the DSO agree on an exchange profile at the TSO-DSO interconnection and each of them is responsible to keep that schedule, while avoiding congestions in their own grid, by making use of the flexible resources located in their network.

Figure 6.21 shows the general scheme of the pilot. The TSO and the DSO agree on the exchange profile at the interconnection, and the DSO, based on the measurements received from smart metres and other grid monitoring devices, launches the local market, which is implemented within the control system. The aggregators (also called commercial market parties, CMPs) optimize the operation of the DER and send bids to the market. Once the result of market clearing is received, the aggregators control the DER, which provide the required flexibility.

The local market is organized to encourage the flexible resources to provide its flexibility and the market pays for the activated flexibility services with marginal price policies. The local market clearing process accepts the lower price offered for the needed flexibility for the grid.

Therefore, the DSO adds two new functionalities to its current activities. On the one hand, it defines, together with the TSO, the exchange profile at the TSO-DSO

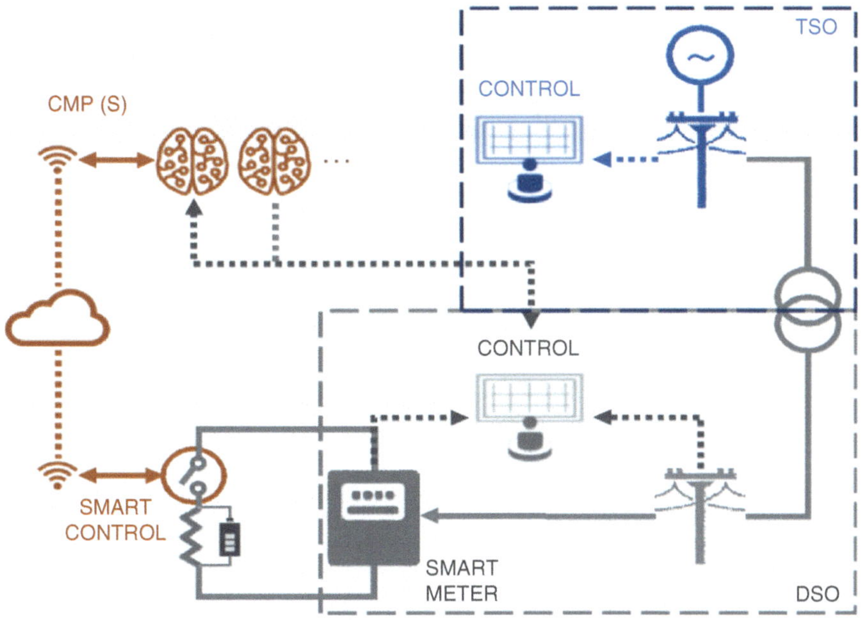

Fig. 6.21 General architecture of the Spanish pilot

interconnection, while, on the other, it performs the local market operator role. The execution of the pilot uncovered the potential of these new two roles by the DSO, enriching the discussions at different stakeholders' forums as potential future roles for these operators.

6.4.2 *The Pilot Project*

The pilot exploited the flexibility available in radio base stations for mobile phone communications. These radio base stations are equipped with backup batteries to guarantee the communications service in the rare event of a failure in the electricity supply. Due to regulatory requirements, these batteries must be able to keep the communications service for, at least, 2 hours after the blackout. Therefore, they can be disconnected from the grid on purpose to provide AS for the operation of the distribution grid.

The pilot was executed in Barcelona and involved 18 base stations, with an average flexibility of about 10 kW. In order to avoid disturbances in the communications service during the execution of the pilot, the base stations were spread along five primary substations, as shown in Fig. 6.22. Given the geographical closeness of the five substations involved in the pilot, they were assumed to represent a single TSO-DSO interconnection point.

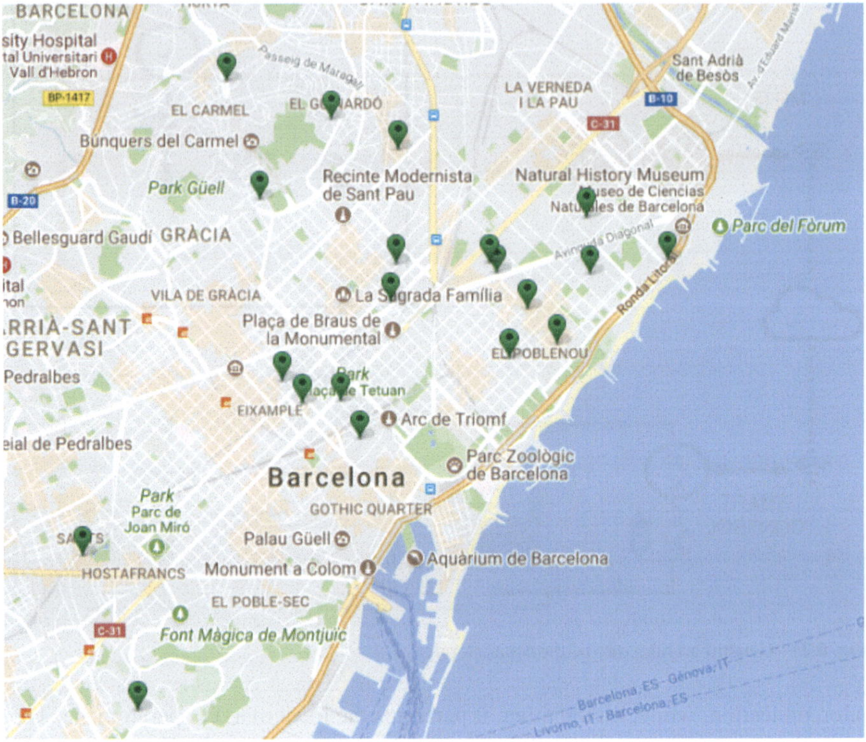

Fig. 6.22 Location of the base stations in Barcelona

During the pilot, the local market was organized through a software platform (Fig. 6.23) developed to receive the offers of flexibility and perform the clearing process. In this case, the DSO was also the operator of the local market, which was used to acquire the required AS both to respect the scheduled exchange profile at the TSO-DSO interconnection and to solve local congestions in the distribution grid.

Therefore, the DSO is responsible for matching a scheduled active power profile at a virtual TSO-DSO interconnection point. This profile, with 24-hour horizon and 15-minute resolution, was generated on a day-ahead basis and given as input data for the execution of balancing services the next day.

In order to solve congestions in the distribution grid, the distribution network downstream the TSO-DSO interconnection point was modelled with enough detail so that balancing at the interconnection and congestions and activated flexibilities were observable. Due to the robustness of the distribution grid in the selected area, there were no real congestions in the distribution grid, so a simulated grid model was used instead. Taking the real grid model as a basis, additional DER and consumption units were included in the simulated model with a twofold objective: to have congestions appearing in the distribution grid and to allow the aggregator

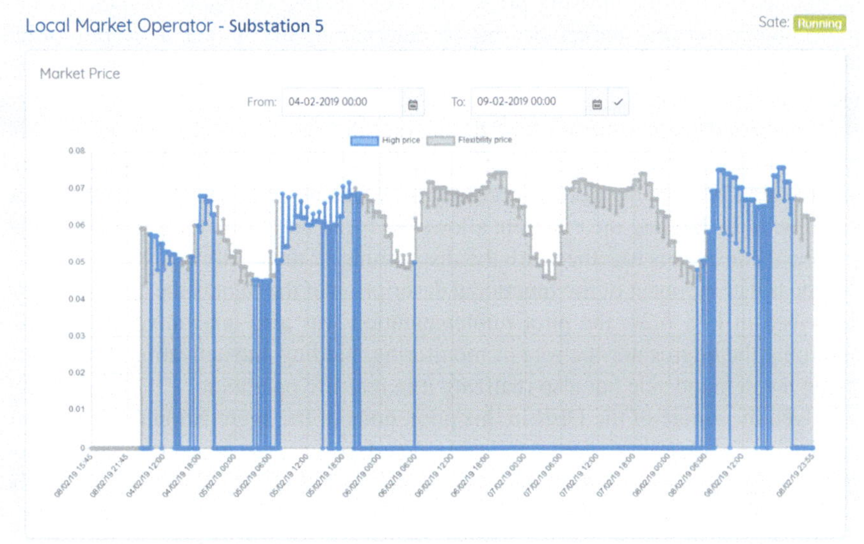

Fig. 6.23 Local market operator page

activating the real base stations to compete against additional simulated aggregators to provide the flexibility required by the DSO.

Moreover, this approach also allowed using simulated market prices to avoid real monetary exchange (which are not permitted by law yet), while focusing on the main objective to evaluate the viability, risks and feasibility or this system.

6.4.3 *Functional Description*

As described in the previous section, the aim of the local market is both to gather balancing services and to solve local congestions at distribution level. To this aim, the mathematical model used to obtain market clearing results combines technical network constraints with flexibility bids. This model is one of the main innovations of this pilot, i.e. the modification of a classical optimal power flow problem (OPF) with the allocation of the flexibility volumes according to the minimum total flexibility required and their activation costs.

When insufficient flexibility is made available by aggregators or technical constraints cause infeasibility on the required active power exchanged at the TSO-DSO interconnection point, the optimization model requests an additional flexibility volume that the DSO should provide to comply with the balancing service.

The consumption is monitored at the TSO-DSO interconnection point, both to assess imbalances and to forecast grid status. This monitored data is used as an input in the modified OPF model to clear the local market and to obtain the flexibility to be dispatched and the clearing price. The local market operator (the DSO in this pilot) implements the market clearing by determining the optimal activation of bid blocks among all aggregators and the clearing price is the most expensive matched bid (marginal pricing).

The second main innovation of the market defined in the Spanish pilot is the execution time, which was set to 5 minutes, to get it close to real-time operation and to be able to provide the needed accuracy to balancing and control. The OPF, combined with the bidding information, allows the local market operator to evaluate the technical constraints together with the dispatching of the flexibility.

The last component of the functional description of the pilot is the aggregator. As described in Fig. 6.24, the pilot implementation was also innovative by not only assigning the aggregator the role of monitoring, bidding and activating the flexibility on the demand side but also realizing it in real-life operation.

Given the nature of the DER in this pilot, none of the aggregation models developed within the project were applicable (see Chap. 3), so a specific aggregation model was developed in the pilot. Moreover, the bidding strategy should also be defined within the aggregation model. For purpose of the pilot, the bidding strategy

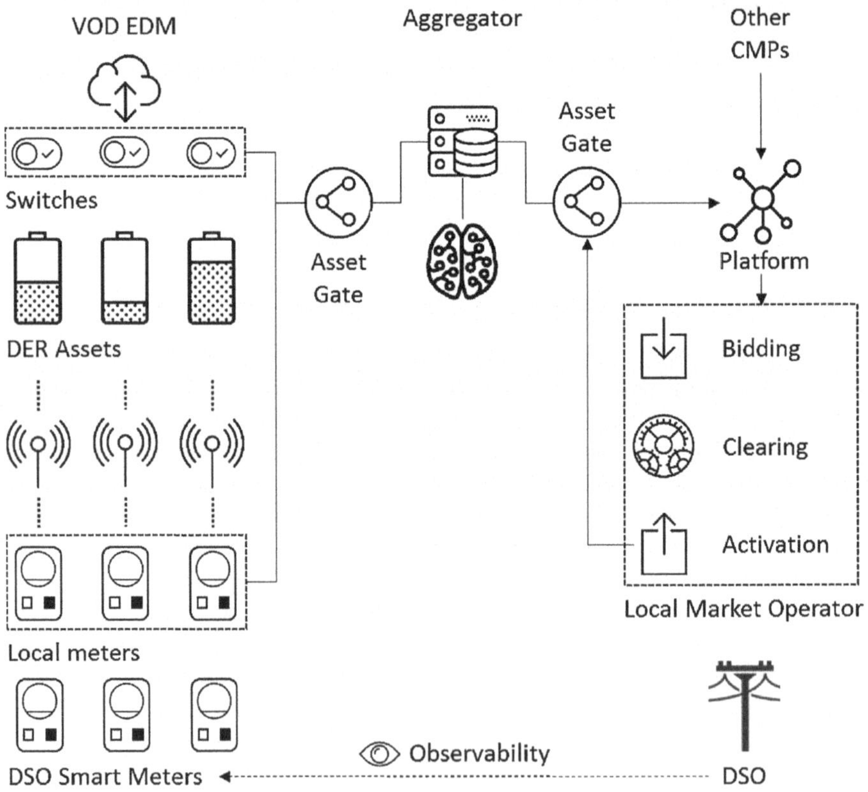

Fig. 6.24 Aggregator process

Fig. 6.25 Commercial market party

is based on the difference between day-ahead and intraday market prices, which allow the aggregator to estimate the need for upward and downward balancing and, hence, to build two bidding curves, with up- and down-regulation. The rebound effect is also considered when deciding the asset to be activated.

Figure 6.25 represents an example of the amount of flexibility provided by an aggregator, where the grey line is the total flexibility available, the light blue is the expected consumption and the dark blue is the real consumption.

6.4.4 Results

During the testing phase, which was divided into 3 weeks, the pilot was kept in operation continuously. The DSO calculated the flexibility required, the aggregators estimated the available flexibility in the radio base stations and sent the bids to the local market, the market was cleared, and the clearing results communicated and executed as required.

For each week during the testing phase, the day-ahead and the last intraday prices were gathered. As shown in Fig. 6.26, the intraday prices swing around the day ahead for different hours in the day, mainly due to expected changes in demand and RES generation forecast.

The aim of the tests was to check whether the DSO could meet the goals of maintaining the scheduled profile while avoiding congestions in the distribution grid. However, the flexibility used in the pilot was only provided by reducing demand and, thus, no increase in consumption was done. Therefore, the DSO could contract

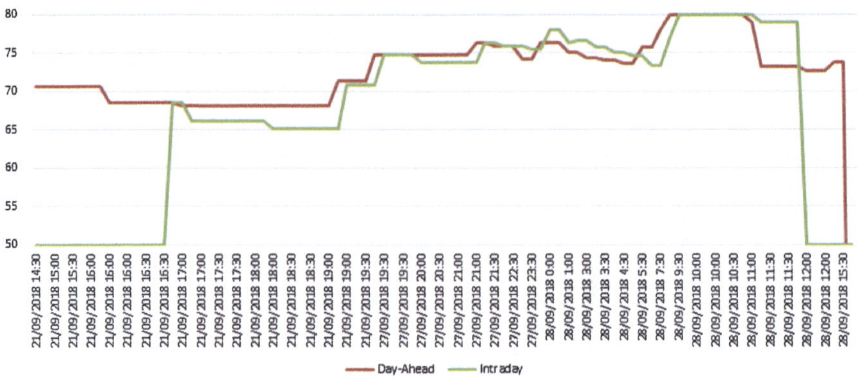

Fig. 6.26 Example of wholesale electricity prices

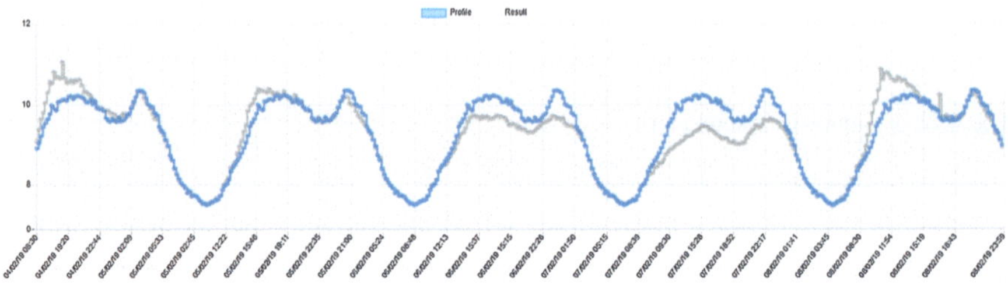

Fig. 6.27 Balancing of Substation 3, during the execution between 4 and 8 of February 2019

upward balancing (demand reduction), but not downward balancing (demand increase or generation reduction), because the types of DER to provide it were not included in the pilot.

The result of the demand gathering by the DSO for balancing purposes is depicted in Fig. 6.27, where the blue line represents the scheduled profile and the grey line is the actual exchange. Likewise, Fig. 6.28 presents the flexibility required by the market (dark blue) and the flexibility made available by aggregators (light blue). As shown in the figures, when there is not enough flexibility available to meet all the requirements by the DSO (e.g. on February 4 in the morning), the dark blue line and the light blue line get together in Fig. 6.28 and the grey line is above the blue line in Fig. 6.27. On the contrary, when the DSO needs downward balancing (e.g. in February 6), there is no available flexibility, so the dark blue line represents zero in Fig. 6.28 and the grey line is below the blue line in Fig. 6.27. In the rest of the cases, that is, when there is enough flexibility available to meet DSO's needs, the DSO meets the scheduled profile and, hence, the grey and the blue lines in get together.

The pilot was able to meet the scheduled exchange profile (except when there was not enough available flexibility or the demand had to be increased) not only in Substation 3, but also in the rest of substations involved in the pilot, as shown in Figs. 6.29 and 6.30.

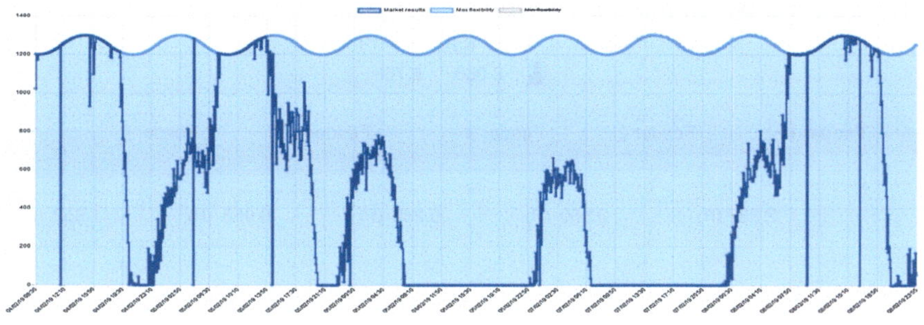

Fig. 6.28 Flexibility of Substation 3, during the execution between 4 and 8 of February 2019

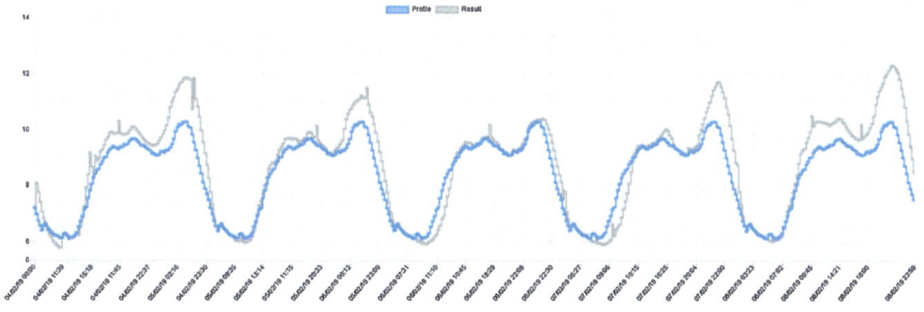

Fig. 6.29 Balancing of Substation 4, during the execution between 4 and 8 of February 2019

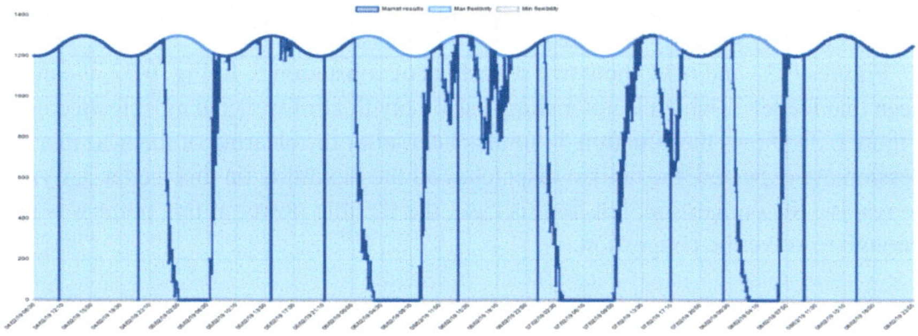

Fig. 6.30 Flexibility of Substation 4, during the execution between 4 and 8 of February 2019

Regarding the avoidance of congestions, Fig. 6.31 shows a screenshot of the status below Substation 4. In this screenshot, there is a real base station providing 5 kW of flexibility (in feeder 3) and other virtual stations and lines are coloured to show the loading rate of the lines. In each DER, the number above represents the baseline consumption, the number in the middle the activated flexibility and the number below the actual power consumption, all of them in kW.

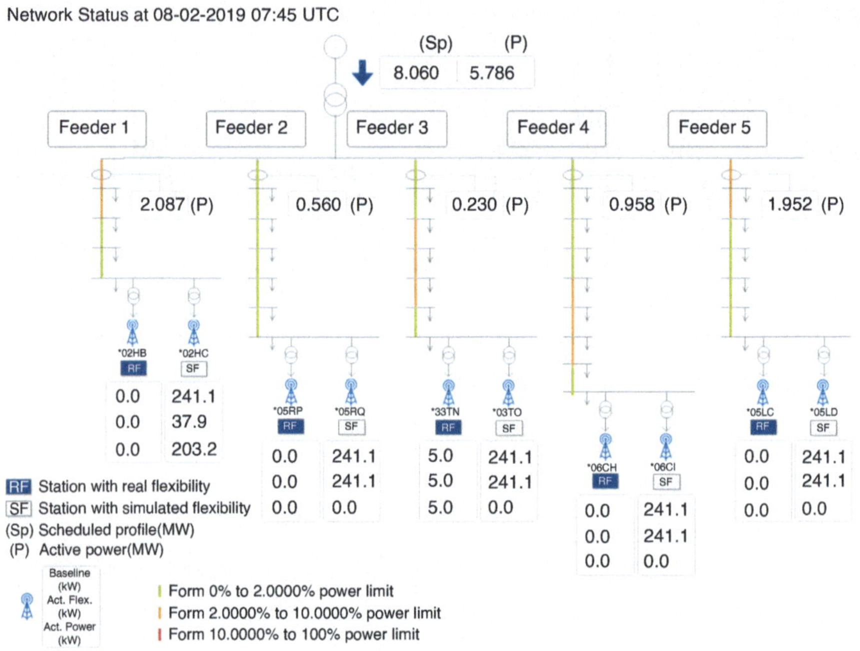

Fig. 6.31 Substation 4, market dispatch at 08-02-2019 07:45 UTC

In this case, all the available flexibility in feeders 2, 3, 4 and 5 has been dispatched (as active consumption in all of them is zero), while the flexibility in feeder 1 has been only partially activated.

Figures 6.32 and 6.33 show two snapshots of substation 5. In Fig. 6.32, it can be seen that feeder 3, which is not making use of any flexibility, is the most loaded one. Figure 6.33 shows the situation 5 minutes later, after the clearing of the next market session. As expected, the market dispatches all the flexibility on this feeder 3, trying to resolve the congestion, but, in this case, the flexible power at this point was not enough to solve the congestion.

6.4.5 *Lessons Learnt*

The penetration of RES and the new technologies being installed in the power systems require updates and the accommodation of new management schemes. This pilot tested some improvements to give more flexibility to the network in order to overcome the system security challenges derived from the energy transition.

In terms of the DSO, the pilot has tested new roles and responsibilities for the DSO and the grid exploitation. These new experiences give new lessons learned to apply in the future to the grid. The congestion management algorithms combined

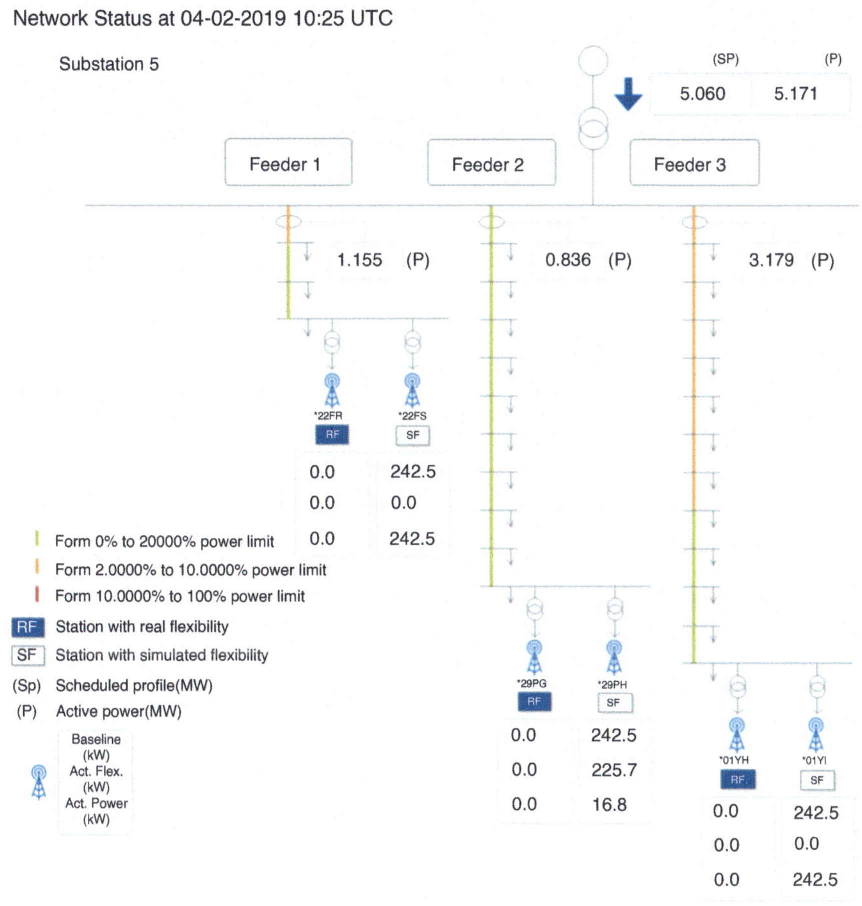

Fig. 6.32 Substation 4, market dispatch at 08-02-2019 07:45 UTC

with bidding clearing models give an alternative method to solve the congestions into the grid versus other solutions such as grid reconfiguration.

As explained, the scheme implemented in the pilot assigns part of the balancing responsibility to the DSO. This new responsibility requires the existence of more available flexibility in order to avoid situations where there is not enough flexibility offered in the market to get the balance. That makes essential a DSO monitoring system at consumer level in low voltage, and the smart meters need to be configured to allow these new functionalities. The observability of the network becomes essential to make better decisions. Compared with the pilot, the current metering systems have more relaxed time requirements than the operation requirements for flexibility management.

Also, this pilot allowed the DSO to run a "quasi-real time" market with technical constraints, in contrast to other approaches that solve technical restrictions after

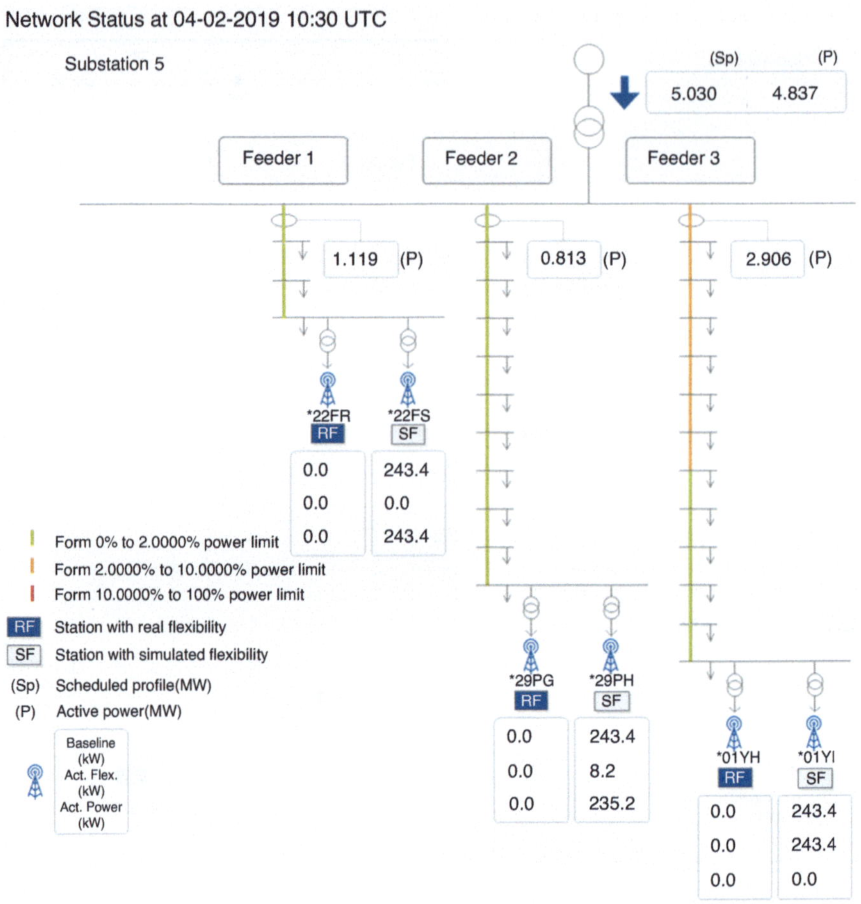

Fig. 6.33 Screenshot of the Substation 5 network, 04-02-2019 10:30

clearing the market for balancing. At the same time, the DSO has tested an optimization algorithm with two objectives: comply with the balance and avoid the congestions.

The aggregator has experimented and applied several lessons learned. First, the use of communication protocols has proven to be useful to allow fast, transparent and non-discriminatory access to market information. From the functional point of view, the pilot served to test the design, implementation and running time errors and experiences under real-world conditions, dealing with communication issues, standardization problems, asset constraints and the sort. In spite of some implementation issues, the aggregator obtained many of the expected results in terms of flexibility and stability by participating in each of the test periods performed over the active lifetime of the pilot.

One important consideration with respect to the mathematical model developed for the bidding strategies is that there is significant space for exploring different techniques, especially in the context of enhanced flexibility (more base stations and/or more flexible batteries) and also aggregation of various types of assets, thereby interlinking their flexibilities to form complex bidding strategies.

This pilot also enabled DER owner to fully run the remote battery test function in full, due to a failure in sending the stop signal by the CMP, hence demonstrating the automatic reinstatement of the rectifiers to normal operation mode once the battery reached the bottom of safety voltage.

The DER owner also faced some issues to deploy all the expected stations. First, some power systems were not compatible with flexibility services and the installed storage systems did not have the correct functionalities. Second, working in a live environment always requires integrating external planning constraints, such as customer service and third parties' constraints, into the field operation. These constraints include the opening time for site access in premises of commercial landlords and residential areas, network freezing periods to maximize the communications network stability in peak periods (e.g. Christmas time or Mobile World Congress), permits for crane setup when needed for equipment replacement on city centre rooftops, etc.

However, the most important lesson extracted from the pilot is that none of the activities had any impact on DER owner customer service. From the DER owner point of view, this is a positive business case and similar players in the market could represent an opportunity for the commercial deployment of aggregation as a third-party service. The Spanish pilot demonstrated that, in good grid conditions, the unused available capacity backup aggregated from base stations can be reused by the DSO for congestion management, eventually avoiding costly ignition of thermal power plants. The replicability potential of the pilot is huge, as, e.g. the DER owner involved in the pilot would be able to trade more than 250 MW of dispatchable load if all their base stations across Europe were aggregated.

This pilot was a good proof of concept for the estimation of the potential of flexible consumers in the provision of ancillary services in Europe. The pilot's intelligence base is extracted from the models devised in several contributions from the partners involved in the Spanish pilot. To validate the technologies developed and incorporated, the Spanish pilot carried out laboratory tests, simulations and real field tests. Throughout these validation phases, the Spanish pilot faced several challenges, most of which were successfully resolved and which will be considered for improvement in future projects.

The demonstration of the benefits envisioned in the Spanish pilot may contribute to a regulatory change in the next years to help unlock the value of small and multiple-site infrastructure assets owned by telecom operators (and other similar DER), while contributing to the social welfare of European citizens.

6.5 Conclusions

Whenever new concepts are proposed, it is important not only to assess their theoretical and economic feasibility but also to verify that they can be really implemented in field.

The three pilots deployed within SmartNet demonstrated the technological feasibility of the concepts proposed, by implementing different TSO-DSO coordination schemes and market structures, by focusing on different aspects of the procurement of AS and by using different types of DER. In general, the three pilots were successful in procuring AS from DER, and the minor issues resulting from the real-life implementation were very useful lessons learnt for the replicability of the solutions proposed.

The Italian pilot demonstrated the feasibility of having a real-time aggregation of information of the units located in the distribution grid and the communication of such information between the TSO and the DSO. It also demonstrated that DER can provide AS for the TSO, although the capability of DER to regulate voltage at transmission level is significantly lower than the potential of big power plants, and that their response time is not in line with the present requirements of the automatic frequency restoration reserve process but that it is with the requirements of the replacement reserve process.

The Danish pilot demonstrated the feasibility of using penalty signals to modify the consumption pattern of indoor swimming pools, which can be used to provide AS to the TSO or the DSO. It was also discovered the importance of having a strong communication network to allow the system to work, which reoriented the focus of follow-up work to wastewater treatment plants, which are located in urban environments with better communication infrastructure.

The Spanish pilot demonstrated the feasibility of DSO-managed local markets to maintain a scheduled exchange programme at the TSO-DSO interconnection while avoiding congestions in the distribution grid. Moreover, it opened the door for radio base stations to provide flexibility to aggregators, which can then offer AS to the DSO, without any impact on their core business of providing communications service. The use of standard protocols and to ensure a proper vendor management also proved to be key for the success of the pilot.

References

1. Terna, Grid development plan (2019). Available at: http://download.terna.it/terna/0000/1188/36.PDF. Accessed 19 June 2019
2. SmartNet Project: Deliverable D5.1 (17 June 2019). Available at: http://smartnet-project.eu/wp-content/uploads/2019/06/D5.1.pdf. Accessed 19 June 2019
3. EnergiNet, Annual report 2018: Secure supplies are shared supplies. Short version (2019). [Online]. Available at: https://en.energinet.dk/-/media/A81882BF3754406D8ADBC258CF5E2D04.pdf. Accessed 19 June 2019

4. SmartNet Project: Deliverable D5.2 (1 June 2019). [Online]. Available at: http://smartnet-project.eu/wp-content/uploads/2019/05/D5.2.pdf. Accessed 19 June 2019
5. H. Madsen, J. Parvizi, R. Halvgaard, L.E. Sokoler, J.B. Jørgensen, L.H. Hansen, K.B. Hilger, Control of electricity loads in future electric energy systems, in *Handbook of Clean Energy Systems*, (Wiley, Chichester, 2015)
6. Annex 67: Energy Flexible Buildings (Nov 2017). [Online]. Available at: http://www.annex67.org/. Accessed 19 June 2019
7. R. Halvgaard, N.K. Poulsen, H. Madsen, J.B. Jørgensen, Economic model predictive control for building climate control in a smart grid, in *EEE PES Innovative Smart Grid Technologies (ISGT)* (2012)
8. O. Corradi, H. Ochsenfeld, Integration of fluctuating energy by electricity price control, Master's Thesis, Technical University Denmark, 2011
9. O. Corradi, H. Ochsenfeld, H. Madsen, P. Pinson, Controlling electricity consumption by forecasting its response to varying prices. IEEE Trans. Power Syst. **28**(1), 421–429 (2013)
10. R.G. Junker, A. Ghasem Azar, R.A. Lopes, K.B. Lindberg, G. Reynders, R. Relan, H. Madsen, Characterizing the energy flexibility of buildings and districts. Appl. Energy **225**, 175–182 (2018)
11. H. Madsen, J. Holst, Estimation of continuous-time models for the heat dynamics of a building. Energy Build. **22**, 67–79 (1995)
12. P. Bacher, H. Madsen, Identifying suitable models for the heat dynamics of buildings. Energy Build. **43**(7), 1511–1522 (2011)
13. H. Madsen, *Time Series Analysis* (Chapman and Hall/CRC, Boca Raton, 2007)
14. R. Juhl, J. Kloppenborg Møller, H. Madsen, ctsmr-Continuous time stochastic modeling in R. arXiv preprint arXiv:1606.00242 (2016)
15. N.R. Kristensen, H. Madsen, S.B. Jørgensen, Parameter estimation in stochastic grey-box models. Automatica **40**(2), 225–237 (2004)
16. SmartNet Project: Deliverable D5.3 (16 Apr 2019). Available at: http://smartnet-project.eu/wp-content/uploads/2019/04/D5.3_20190415_V1.0.pdf. Accessed 19 June 2019
17. ENTSO-E, Ten-year network development plan 2016. Executive report (20 Dec 2016). Available at: https://tyndp.entsoe.eu/exec-report/. Accessed 19 June 2019
18. Red Eléctrica de España, Spanish Peninsula – Electricity demand tracking in real time. Available at: https://demanda.ree.es/visiona/peninsula/demanda/total/2019-01-26. Accessed 25 June 2019

Chapter 7
Regulatory Frameworks for Enabling Distributed Energy Resource Participation in Smart Grids

Ivana Kockar, Dario Siface, and Andrei Morch

7.1 Introduction

To realise low-carbon electricity networks, we need to increase levels of renewable energy resource (RES) connections, which, due to intermittency and variability of such resources, call for additional instruments to facilitate their integration. Because of a need to balance demand and supply at any instant in time, one of the essential issues when operating electricity systems with high levels of RESs is a support for system balancing. Although uncertainty and a need for balancing have always been considered and system operators always kept a level of reserve to account for unexpected events, high levels of RESs bring this problem to another level and require solutions that will rely not only on the provision of a reserve from available generation but also other, less expensive solutions that include provision of flexibility by resources such as demand side response or utilisation of energy storage.

RESs can be introduced at all voltage levels, but are often connected at distribution networks, where they have introduced a paradigm shift in operation of these networks which are thus becoming active. Injections of active power may cause changes in directions of power flows and, then, other issues such as voltage rise at certain network points, as well as network congestion. A number of solutions have started to emerge to address this issues, from Active Network Management [1] applied in the UK to curtail generation output to more complex approaches to

I. Kockar (✉)
University of Strathclyde, Glasgow, UK
e-mail: ivana.kockar@strath.ac.uk

D. Siface
RSE S.p.A., Milan, Italy

A. Morch
SINTEF Energy Research, Trondheim, Norway

G. Migliavacca (ed.), *TSO-DSO Interactions and Ancillary Services in Electricity Transmission and Distribution Networks*,
https://doi.org/10.1007/978-3-030-29203-4_7

provide flexibility by active demand side participation of smaller customers (e.g. commercial or domestic) and energy storage (including electric vehicles). These distributed devices, including RESs connected at distribution networks, are typically referred to as distributed energy resources (DERs).

The changes in the operation of networks due to integration of RESs require a development and implementation of new operational and market tools and arrangements. These will help provide flexibility services needed to integrate energy from the renewable resources, as well as enable electrification of other sectors, such as transport or provision of heat energy. This, however, needs to be done in a cost-effective way that also seeks to avoid, reduce or delay infrastructure reinforcement.

Furthermore, integration of renewable resources and other technologies that can provide flexibility has also to be carried out considering the overall EU and national energy regulation frameworks. To achieve this, it is important to develop approaches that help integration of flexibility services provided by DERs into energy markets, so that they can compete at providing ancillary services. While trading of ancillary services at transmission level is quite well defined and managed by transmission system operators (TSOs), and DERs can take advantage of these markets, provision of such services at the distribution level has been less well defined so far. There is no doubt that DSOs would benefit from being able to utilise flexibility to resolve issues appearing at the distribution level, due both to RESs integration and ancillary services trading offered by DERs to TSOs. As recently proposed by Clean Energy for All Europeans package [2, 3] and other European regulations, distribution system operators (DSOs) are encouraged to procure flexibility offered by DERs to manage congestion on their systems and for other non-frequency ancillary services that include voltage control. However, procurement of these services will have to be allowed and regulated by national regulation.

Enabling of DERs participation in the ancillary services markets requires additional coordination between TSOs and DSOs, which was the subject of the SmatNet project, with the five coordination schemes (CSs) outlined in Chap. 3. Analysis of these coordination schemes was carried out based on the results from the simulator, which can contribute towards developing an active system management toolbox [4] that calls for different types of solutions to help with managing ancillary services markets for congestion management and balancing.

As discussed in [4], these solutions include:

(i) Technical solutions (such as management of reactive power flows, grid configuration, etc.)
(ii) Tariff solutions that can be used to entice customers to provide response based on time, location, etc.
(iii) Market solutions that will enable customers to sell their flexibility via organised markets
(iv) Connection arrangements with certain users defining when and how they will be obliged to provide certain services whenever needed
(v) Rule-based approaches to curtailments that may be necessary as a last resort to secure system operation

Although the simulator developed in the SmartNet was formulated prior to the publication of [4], and thus was not following the above toolbox requirements, it nevertheless can be used as a basis for evaluation of some of the above solutions.

The solutions provided by the SmartNet are also in line with the requirement that a successful transition towards these new energy systems has to adequately support integration of DERs and their flexibility services into the congestion and balancing markets while maintaining secure system operation, but also supporting current and emerging European Commission (EC) and national regulation. The aim of this chapter is to provide insight into outcomes of two main aspects regarding regulatory analyses that are necessary to consider when developing new TSO-DSO coordination schemes. These include: (i) how the proposed solutions fit within the current and emerging EC and national regulation, road maps and position papers and (ii) what are the important aspects that need to be considered when developing and implementing these schemes. Although the presented discussion will be based on the analysis of coordination schemes presented in the SmartNet project, the drawn conclusions are valid and applicable to other similar approaches and could be used as a basis for further developments that could be more suitable for a particular power system network.

First, This chapter briefly outlines a policy framework that has been shaping electricity industry and, in particular, latest developments that are enabling realisation of low-carbon electricity networks with high levels of renewable generation. It also outlines latest changes that are putting customers at the centre of energy transition, thereby affecting TSO and DSO operation. This is opening a question of their better coordination, and, thus, a need for solutions proposed by, and evaluated in, the SmartNet project. This chapter also discusses changes that are necessary in the DSO operation to enable better DER integration and discusses results from implementation of five coordination schemes analysed in the SmartNet project. Finally, it discusses general issues that need to be considered when deciding on rules and regulation that will enable shift to systems that will require increased flexibility provided by different technologies and customers.

7.2 Main Regulatory Frameworks and Stakeholders' Positions Affecting TSO-DSO Coordination

Main pillars of the European energy and climate change polices include integration and use of renewable generation so to reduce greenhouse gas emissions and increasing energy efficiency, while ensuring secure and affordable energy supply. To achieve this, a number of frameworks have been developed with an aim to introduce, reform and coordinate electricity markets. The First Energy Package/Directive aimed at introducing competition, with the objective of separating or unbundling former energy monopolies and ensuring the distinction between regulated and non-regulated activities for electricity [5], was adopted in 1996. In June 2003, the Second Directive sought to reinforce separation whereby energy transmission networks had

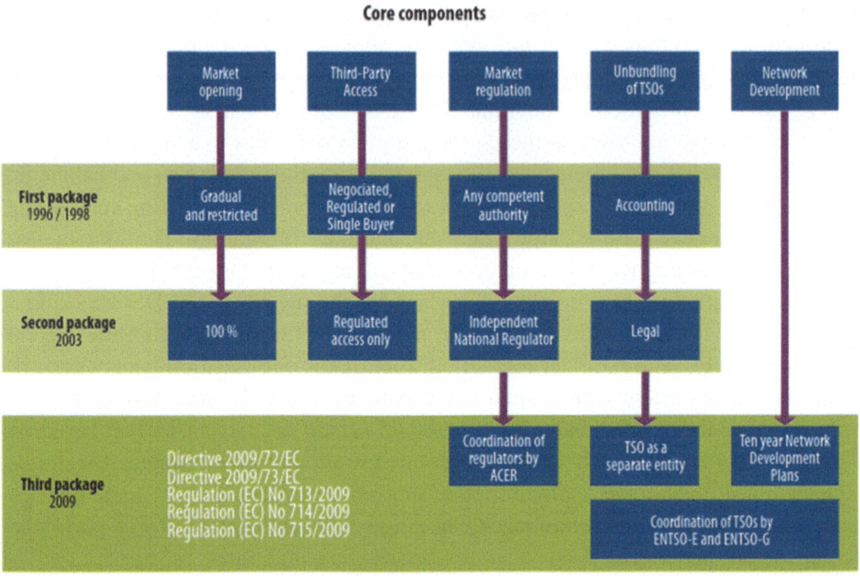

Fig. 7.1 Development of the three energy packages [12]

to be run independently from the production and supply of energy [6], while the Third Energy Package of 2009 consists of a number of directives and regulations, including two directives and a regulation for electricity:

(i) Conditions for access to the network for cross-border exchanges in electricity [7]
(ii) Regulation establishing an Agency for the Cooperation of Energy Regulators [8]
(iii) Common rules for the internal market in electricity [9]

Integration of gas and electricity networks was also regulated by the Third Energy Package [10, 11]. Development of the three energy packages is shown in Fig. 7.1 [12]. In addition to energy packages, Emission Trading Scheme, which is a cap and trade scheme, was initially introduced in 2005 and is currently in its third phase (2013–2020) [13].

Coordination and harmonisation of energy transition towards secure and affordable energy systems has also led to the Energy Union policy initiative, which was introduced in 2015 [14]. The main objective of the Energy Union is to reinforce and ensure (i) energy security, solidarity and trust; (ii) fully integrated European energy market; (iii) decarbonisation of the overall economy; (iv) reduction of energy demand via energy efficiency measures; and (v) research and innovation.

The latest in the series of EU regulatory frameworks was the Clean Energy for All Europeans package [2, 3] that had three main goals:

- Putting energy efficiency first
- Achieving global leadership in renewable energies
- Providing a fair deal for consumers

To enable customers to participate in electricity markets, the Clean Energy for All Europeans legislative proposals also cover the design of the electricity market, and provide opportunities to design markets that will enable passing of real-time prices to the final customers. These price signals should entice consumers to participate in the provision of flexibility by increasing or decreasing their consumption and thereby helping system operators to run the systems securely and in a cost-effective way.

In addition, government subsidies, as well as a price of PV and other renewable technologies, are decreasing, with an increasing number of customers installing these technologies, thus becoming "prosumers", i.e. at certain times producing more than what they consume and exporting this surplus power into the grid. However, there are still no sufficiently well-defined rules that would enable prosumers to sell their energy into the grid, with additional rules that will enable this being discussed as part of the Clean Energy for All Europeans.

The package also recognises that connection of prosumers, and DERs in general, is typically occurring on distribution networks, which opens additional problems in their operation, but also in the organisation of electricity markets. To that end, the package recognises that local energy communities (and possibly local markets) will provide opportunity to participants at a local level to participate more efficiently in the market. It will also enable synergy of multi-vector energy systems, as flexibility can be provided by other technologies such as storage heaters, water tanks as well as electric vehicles.

However, trading of flexibility and participation of DERs in ancillary services markets at both transmission and distribution levels can cause issues such as voltage violations and congestions on distribution networks. To overcome this, the Clean Energy for All Europeans package also considers allowing DSOs to manage these issues via locally procuring ancillary services that will help resolve them. This, however, has to be aligned with the national regulation, although member states are encouraged to enable market solutions and/or DSO neutrality when managing local network issues. Therefore, the aim of the proposed polices is not to impose a top-down approach in managing electricity markets, but rather a compromise between bottom-up initiatives and top-down steering of the market. Such approach leaves more freedom to national governments and regulators to better adjust solutions to their needs, while seeking to provide a level playing field to all participants; however, there is still a need for national authorities to agree to common trading rules and methodologies to enable cross-border trading.

Finally, smart metering infrastructure, as well as availability of aggregators that can help small customers to participate in the market, will be crucial to realise these low-carbon energy systems. This further opens the question and a need for a regulation of data exchanges and data protection, as smart metres will generate a range of energy data carrying high commercial value. New ways of operation and coordination of large number of customers and DERs also require adequate ICT infrastructure that can support both centralised and decentralised operations, as discussed previously in Chap. 4, and in more details in [15].

In addition, coordination between TSOs and DSOs is crucial for the secure operation of their networks, as has been recognised by a number of stakeholders whose positions need to be considered. These key stakeholders are in many ways defined by implementation of the latest changes in the European legislation related to the internal gas and electricity markets, i.e. the Third Energy Package (entered into force in 2009). The package established National Regulatory Authority (NRA) for each member state and a common Agency for the Cooperation of Energy Regulators (ACER) [8]. Following the same process, the European Network of Transmission System Operators (ENTSO-E) was established in 2008 as a common body representing European TSOs. Note that ENTSO-E in fact has a twofold functional role. On one hand, it operates as an organisation which represents interest of the Europeans TSOs, and on another, it acts as regulatory body which develops the network codes (guidelines). Moreover, it is necessary to consider positions of organisations that represent stakeholders working with different aspects of regulation and standardisation, industrial associations such as European Distribution System Operators (EDSO), the European Federation of Local Energy Companies (CEDEC), European independent distribution companies of gas and electricity (GEODE), Union of the Electricity Industry – Eurelectric, and Council of European Energy Regulators (CEER), which have published documents that reflect their positions and needs, as well as their views on TSO-DSO cooperation [4, 16–20].

The long-term visions, which also reflect regulation, are often presented in road map documents. For example, the latest ETIP SNET Vision 2050 [21] provides a comprehensive view of a number of stakeholders on integration of entire energy systems that encompass energy vectors. This vision builds on the previous road maps, such as European Technology Platform (ETP) for Smart Grids (SG) Vision for 2020, published in 2006 [22].

In the SmartNet project, focus has been on the third aspect of the Energy for All Europeans package that seeks to put consumers at the heart of the energy transition and enable their participation in the market. This means that integration of RES will be supported by the active participation of customers who can provide valuable flexibility by changing consumption depending on the system needs and participate in providing ancillary services at both transmission and distribution levels. However, this may require changes in the operation of DSOs and the roles they play, as well as new approaches to interactions between TSOs and DSOs. Comparative analysis of concepts and solutions investigated in the SmartNet against a set of regulatory documents, road maps and other documents published by stakeholders has been presented in [23], while learnings from the SmartNet project and policy recommendations have been detailed in [24, 25], respectively.

7.3 A Change of Perspective in Distribution Network Operation

In the previous chapters, all the TSO-DSO coordination schemes (CSs) that have been assessed within the SmartNet project have been presented in detail. It has become apparent that the main difference among all the schemes lies in the role that

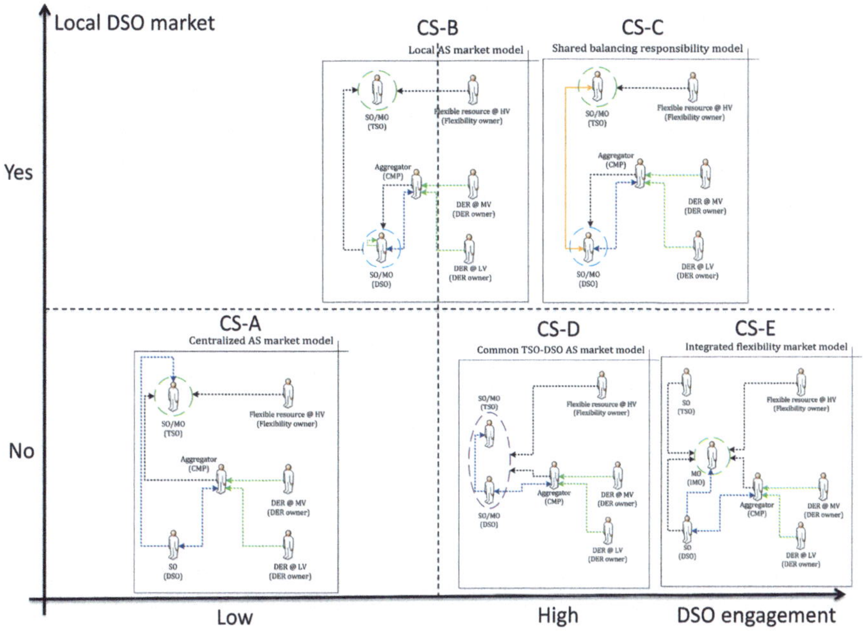

Fig. 7.2 Graphical representation of the main characteristics of the different TSO-DSO coordination schemes assessed in the SmartNet project

the DSOs have under new system operation paradigm. As graphically described in Fig. 7.2, the main differences between the evaluated CSs are:

(i) The possibility for DSOs to have access, by means of a market mechanism, to flexibility resources to be used for the operation of the local network, which in Fig. 7.2 is represented on the vertical axis – local DSO market
(ii) Levels of DSO involvement in the operation of the market, represented on the horizontal axis – DSO engagement

The coordination schemes evaluated in the SmartNet project are presented in details in Chap. 3, with market design explained in Chaps. 3 and 5, and implementation details explained in Chaps. 4, 6, and 7. These coordination schemes are as follows:

- *Coordination Scheme A* (CS-A), where the DSOs have only the assignment of prequalifying the resources connected to the local networks. This prequalification needs to ensure that DERs can be activated to provide ancillary services (AS) when called upon by TSO only if such an operation will not cause distribution network issues, e.g. congestion. It is assumed within the SmartNet project that CS-A will be common TSO-DSO coordination scheme in Europe in 2030.

- *Coordination Scheme B* (CS-B), in which the DSOs operate a local market in order to acquire flexibility resources for the congestion management of their local networks.
- *Coordination Scheme C* (CS-C), where the DSOs operate a local market in order to acquire flexibility resources both for the congestion management and for the balancing of their local networks.
- *Coordination Scheme D* (CS-D), where TSOs and DSOs operate together a global ancillary services market in order to acquire flexibility resources they need for the operation of their networks.
- *Coordination Scheme E* (CS-D), where an integrated flexibility market is introduced and in which system operators and commercial market parties compete for flexibility resources. Note that due to mathematical complexity, this scheme has not been assessed in details as the other four.

The shown path of change in the role of the DSOs from CS-A to CS-B, CS-C and finally CS-D and CS-E, as depicted in Fig. 7.2, is strictly dependent on the underlying assumptions made in the SmartNet project. It is very likely that a different set of assumptions may lead to different CSs, as well as to different paths of their evolvement. In any case, the inclusion of flexibility resources connected to the distribution networks requires development of coordination schemes that will govern the TSOs and DSOs roles, as well as their interactions. This will be essential for the future power system operation and will become ever more important as increased integration of RESs, and in turn DERs, that is not adequately managed and coordinated may significantly increase a need for expensive network infrastructure reinforcements. Currently, European regulation mainly considers only the remuneration of investment in network expansion, while DSOs have no access yet to flexibility resources in their local networks (we do not consider generation curtailment and load shedding as “flexibility resources”). At the moment, in addition to generation curtailment and load shedding, the only course of action available to DSOs is a possibility to resolve network congestion via reconfiguration. Therefore, the present policies governing distribution network operation are mainly limited to the investments in network reinforcements, sometimes oversizing the network, which effectively is equivalent to a so-called “fit-and-forget” policy.

This approach, however, is not always economically (nor environmentally) efficient and calls for complete change in DSOs roles and responsibilities. These changes need to enable DSOs to address emerging issues related to DER integration and their participation in the ancillary services markets at both transmission and distribution level networks. The results of the SmartNet project not only provide technical and economic analysis of the needed changes in the system and market operations, but also seek to evaluate regulatory framework that is needed to enable and support such changes.

The following section of this chapter outlines the experiences gained from the practical implementation of the evaluated coordination schemes. Some of the results are closely dependent on the assumptions made in the SmartNet project, while others are independent and more general.

As discussed in Sect. 7.2, a strong cooperation between TSOs and DSOs is essential, as also acknowledged by position papers and documents published by various stakeholders and associations, e.g. [4, 16–20].

7.4 Assessment of the Different Coordination Schemes

This section presents analysis of outcomes from the implementation of the five coordination schemes assessed within the SmartNet project. In addition to highlighting advantages and disadvantages for each of the schemes, this section also draws some regulation guidance and recommendations from the main results of the project. As mentioned above, complete description of the coordination schemes considered by SmartNet can be found in Chaps. 3, 5, and 6 as well as in [26].

The assessment of the schemes has been performed by means of numerical simulations for each of the three national scenarios for year 2030, i.e. one for Italy, one for Denmark and one for Spain. In addition, each of the three national pilots [27–29], one for each of the mentioned countries, implemented different coordination scheme, which are presented and discussed in detail in Chaps. 5 and 6, as well as in [27–30]. Coordination schemes have also been tested in the laboratory environment as detailed in [31].

7.4.1 CS-A: Centralised Ancillary Services Market Model

The first coordination scheme presented contemplates a market model in which the ancillary services market (ASM) operated by the TSO considers resources connected at both transmission and distribution levels, but it does not take actively into account the local network constraints. The DSOs are involved in the procurement and activation process of AS by the TSO only if a system prequalification scheme is implemented to guarantee that the activation of resources from the distribution grid by the TSO does not cause additional constraints (e.g. congestion) in the local distribution network (DN). In this case, this prequalification of resources at DN level is under DSO's responsibility, and it can be implemented at various degrees: for instance, the DSO has to evaluate the impact of the delivery of a specific service by a certain resource on the grid, and if it violates local grid constraints, the DSO could forbid the delivery of the service by that specific resource.[1]

Whatever the degree of responsibility, in order to allow the DSOs to exercise it, investments in monitoring and control systems (i.e. ICT) are needed, along with a higher level of expertise on DSO side. This is particularly important with respect to

[1] This differs from the technical prequalification, where a certain resource is assessed to make it eligible to deliver a specific service.

smaller DSO. As mentioned above, it means that DSOs should be encouraged and supported in going beyond the present "fit-and-forget" reinforcement policy: for a more efficient operation of the whole system, the local network reinforcement should be compared with the use of flexibility resources, which is made available by investment in ICT and change in regulation. This means that:

(i) Regulations should shift from a CAPEX remuneration perspective to a TOTEX one.
(ii) The investment evaluation should be coordinated between transmission and distribution since it is related to the whole power system operation.
(iii) A long-term planning perspective should be used for the whole system, that is for distribution as well as for transmission.

Otherwise, significant reinforcement can lead to unnecessary and expensive oversizing of the networks so that in the near future DNs may still not experience significant congestions. Allowing expensive network oversizing policies to remain in place may lead to an underestimation of the positive effects of investing in the implementation of monitoring and control system. As a consequence, some DSOs may miss on an opportunity to fully appreciate a value of utilisation of flexibility rather than network reinforcements, which can hinder implementation of other (potentially more efficient) market schemes for decades, even beyond 2030. For example, cost-benefit analysis of the simulation results for the 2030 scenario in Denmark (see Chap. 6 and [30]) indicated that CS-A had the best economic efficiency among the four simulated CSs. This is the consequence of the complete absence of local congestions in the considered Denmark distribution network. Thus, as long as local networks have sufficient capacity (which would typically be oversized), there would be no need to consider other market schemes. However, if the amount of DERs, and in particular DGs, increases, it can be expected that local congestions may begin to occur, and that will be the right time to evaluate which type of investment will be more efficient: network expansion or seeking to implement another market scheme which will also require ICT investments.

7.4.2 CS-B: Local Ancillary Services Market Model

The second coordination scheme presented considers the case of a local market operated by the DSO to select the resources needed to solve the local congestions. Furthermore, the DSO also locally rebalances the system by selecting and compensating the extra activations done so that the TSO does not see the total system imbalance modified. After having cleared the local market, the DSO transfers the remaining flexibility resources to the AS market operated by the TSO. The DSO must also assure that only bids respecting the DSO grid constraints can take part in the AS market.

In practice, this scheme is based on a two-step procedure. The first step is the local market clearing, where all local resources are at disposal of the DSOs, which will choose the least expensive ones taking into account the local network constraints. In the second step, the TSOs solve imbalances and congestions in the whole system using the resources at TN level and the local resources not used by DSOs in the previous step.

The possibility to directly operate a congestion management market offers the DSOs more options to solve grid congestions than the simple grid reconfiguration, as underlined by the experience of Spanish Pilot (Pilot C) [29].

However, CS-B is affected by the following issues:

1. CS-B splits the optimisation that clears the market into two steps, which is likely to result in suboptimal solutions, both from a mathematical and from a technical point of view. For example, consider the case in which the local imbalance caused by the solution of a local congestion is of the opposite sign of the global imbalance: in CS-B, since this information is not at DSO's disposal, it has to solve also the local imbalance by activating other resources, and when it is the turn of the TSO to solve the global imbalance, other activations are needed.
2. Since the local market is limited to the local DN, there may be potential issues with market liquidity if limited resources are connected to DN. It sometimes may also occur that mFRR market cannot resolve the issue and then aFRR, which is usually more expensive, has to be used instead.
3. For the same reasons as above, local DSO market could also be subject to significant potential for the exercise of market power.

As a consequence of these issues, if compared with the operation of market scheme in which the action is centrally coordinated, CS-B is expected to have worse economic performances, which is also revealed by results from the simulations, as discussed in Chaps. 5 and 6, and in [30].

Some possible approaches to remedy each of the three identified in the previous paragraph are outlined below:

For (1): Since the two-step procedure that splits the optimisation that can lead to suboptimal solution is a characteristic peculiar to this coordination scheme, it cannot be completely removed unless the coordination scheme is changed. Nevertheless, in order to reduce the technical inefficiency, some regulatory remedies could be implemented. For instance, to avoid double activations or counteractions, TSOs could be allowed to revoke an accepted bid at the local level, in a way similar to the arrangements adopted in the Italian ancillary services market (Mercato dei Servizi di Dispacciamento – MSD). This market is divided in six different sessions (the first taking place the day before the delivery, with others on the day of the delivery) related to different hours of the day of delivery (MSD1 considers h1 to h24, MSD2 h5 to h24 and so on). At each session the TSO recalculates the demand for the reserve, and then, if bids accepted in the previous sessions are not needed any longer, it revokes them.

For (2) and (3): these are general problems related to all other coordination schemes evaluated in the SmartNet project, although their impact may likely be stronger when the local DNs are treated separately from the rest of the system, as it happens here in CS-B (and also in CS-C, dealt with below, in Sect. 7.4.3). Since it is a common issue, illiquidity of local networks is treated in Sect. 7.5.8.

It should also be taken into account that the local and global markets could be implemented with different clearing frequencies. Then, there is the possibility that a bid that is offered both at local and at transmission levels is accepted twice. In order to avoid this, the setup of the two markets, and in particular of the bidding procedure, should be carefully coordinated by TSO and DSOs, for instance, by implementing a type of "Common Sequenced Market", where a common database of resources is shared between TSO and DSOs without time correlation, so that once a resource has been selected by one operator, it becomes unavailable for others, as also suggested in [16].

7.4.3 CS-C: Shared Balancing Responsibility Market Model

The third coordination scheme considers the case in which the DSOs have complete balancing responsibility for their distribution grids. In this case DSOs operate a local market in order to obtain local resources both for balancing the DN and congestion management. The balancing requirements also include a predefined power exchange schedule at the interconnection node between DN and TN that the DSOs must respect.

CS-C provides the direct access to flexibility resources and, similar to CS-B, offers the DSOs more options for the DN operation than the simple grid reconfigurations. Also, the interaction between different (and often still under development, or in a transition towards electrification) energy carriers (e.g. district heating, hydrogen), as presented in [21], could make available new resources which use could be more easily optimised in a local market scheme. Furthermore, the complete separation between transmission and distribution could, in rare circumstances, bring the positive side effect of preventing that high prices in one area do not spread to the others.

Still, CS-C has the following major drawbacks:

1. DSOs must have a complete control of the networks and of the resources under their responsibility, which requires a significant investment in the ICT infrastructure.
2. The need to respect a scheduled exchange between TN and DN is a very strong constraint. From the mathematical point of view, this means that the clearing solutions found will likely be suboptimal, that is, the economic performances of such a scheme may be poor (as confirmed by simulations [30]) and sometimes may even fail to find a feasible solution. This situation may trigger an imbalance occurring at the interconnection between DN and TN, causing activation of aFRR or unwanted measures (which are more expensive), as shown in simulations [30].

3. This same separation may result in a suboptimal solution, as it is dividing the whole system in smaller subsystems, and thus:
   (a) It may make difficult the "cross usage" of less expensive resources at both levels.
   (b) In the cases when the imbalance in DN has the opposite sign of that in TN, compensation between these two is not possible. Rather, resources must be activated in both markets, which further reduces economic efficiency.
4. Smaller DSO flexibility markets could suffer from illiquid problems, similar to issues that may occur in CS-B. This could cause situations when it may be impossible to clear them, leading to a more frequent, and more expensive, use of aFRR, as shown in simulation results [30].
5. Finally, as in coordination scheme CS-B, here local DSO market may be vulnerable to exercise of the market power by some flexibility resources.

From the regulatory point of view, this coordination scheme is in contrast with the present common vision (embraced also in [4] and Art. 32 in [32]) that balancing should remain a system-wide centralised service procured by the TSOs, or another subject on behalf of the total system, while the DSOs should at most assume responsibility for local regulation (e.g. voltage regulation) and congestion management in distribution.

Finally simulation results presented in [30] show that CS-C may suffer from a significant economic inefficiency when compared to all the other schemes studied within the SmartNet project, confirming the impact of the drawbacks highlighted above.

7.4.4 CS-D: Common TSO-DSO Ancillary Services Market Model

The fourth coordination scheme presented considers the case in which there is a common flexibility market for system operators (SO) in order to minimise total flexibility procurement costs. This scheme can be seen as the evolution of CS-A, in which the market is operated by TSOs and DSOs together, with the DN constraints directly integrated in the market clearing algorithm, so that the outcome of the market clearing does not violate them. Although this scenario is operationally less complex than CS-A, integrating physical grid constraints in the market algorithm on one hand increases the computational cost of the clearing procedure and, on the other, puts the following requirements for each of the DSOs involved:

(i) It has to have the control of its local network and DERs connected, for which adequate investments in ICT support have to be performed.
(ii) It needs to provide the necessary data to the party responsible for the operation of the market.

Since this scheme incorporates all distribution constraints, takes into account the whole system and makes available all the flexibility resources to both TSO and DSOs, it is expected to show high economic performances, even when considering ICT investment cost needed at the DN level to ensure control and data sharing, as explained in items (i) and (ii) above. This expectation is almost fully confirmed by the numerical results from the simulations [30].

However, the above-mentioned expectation of high economic performances can somewhat fell short [30], as in the cases when forecasting error is non-negligible and congestion is low. Then system could take incorrect decisions on the basis of forecasted congestion which does not materialise in real time. This means that resources are activated to solve problems that have been forecasted but have not actually occurred, requiring that further resources have to be counter-activated in real time. Thus, when compared to CS-A (which has been used as the reference for the Cost Benefit Analysis), it may happen that under certain conditions CS-D is more expensive, since the former scheme does not consider distribution constraints and, therefore, cannot be misled by decisions related to DN congestions. This leads to one of the key points related to implementation of coordination scheme CS-D, which is a question on how to deal with forecasting error since it is its only source of economic inefficiency. However, the issue of forecasting errors affects all the possible coordination schemes (even if at a lower degree), and this will be discussed in Sect. 7.5.12 as one of the critical issues that is important for all coordination schemes.

Another important aspect bound to CS-D is the fact that the market should be co-managed by the TSO along with all relevant DSOs. As discussed above in (ii), this could imply that all DSOs have to implement a monitoring/control system within their network. Also, note that the market could be physically under responsibility of a market operator representing all the involved SOs while having the right technical competence to manage clearing algorithms capable of considering network operational constraints.

7.4.5 CS-E: Integrated Flexibility Market Model

The fifth coordination scheme presented considers the case in which both regulated (TSO and DSO) and commercial market parties (CMPs) procure flexibilities in a common market. The mathematical simulation of this scheme is characterised by a high level of mathematical complexity, in particular due to the need to consider game theory elements. For this reason it was neither investigated in details (as the other schemes), nor simulated within the SmartNet project.

Three significant drawbacks for this scheme may be the following:

- Such a scheme may create uncertainty regarding the real amount of congestion and imbalance and, as a consequence, how many resources are needed by TSOs and DSOs, since CMPs can change their output in this market.

- Since TSOs and DSOs compete in equal terms with other CMPs for flexibility, in some scenarios they can no longer be sure even if they will be able to acquire the resources needed.
- The high level of competition between operators could increase market prices.

However, this scheme could prove promising for trading of certain types of AS (for instance, it could perform very well for solving balancing issues) and also to eliminate gate closure issue, while it could also help RES generation to solve their imbalances at a market level, thus reducing the system imbalance. Thus, further investigations of CS-D market schemes are needed.

7.5 General Critical Issues

While each of the above-discussed coordination schemes has its own characteristic, there are also a number of general aspects that are important to consider when developing and implementing any of the proposed schemes. For example, these include aspects related to market design, modelling of the system and network constraints, relationships between different markets, market products, liquidity of markets, importance of forecasting errors, etc. These will be discussed in the following subsections, with details on how they have been addressed in the SmartNet project.

7.5.1 Market Modelling and Timelines

A market model developed and implemented in the SmartNet project sought to enable mechanisms, which would help DERs provide flexibility by trading their energy in the ancillary services markets. The market clearing simulator has been based on a hierarchical design formulated as standard optimisation problem. Market design enables DERs trades with both TSOs and DSOs, and reflects adopted coordination schemes. Market clearing is carried out based on bids submitted by market participants while respecting systems constraints, as well as those of the devices, as discussed in Chaps. 3 and 5.

Due to the nature of the markets and trades, i.e. intraday market for flexibility, as well as technical characteristics of the DERs, the following aspects of the market design and operation are important to consider:

- *Time step* refers to the time granularity considered in the market clearing. At each step, activation decisions for market participants are made, with the assumption that the system and the flexibility assets inside each time step are not changing their behaviour.
- *Time horizon* refers to the overall time period considered for the market operation and clearing. The time horizon can be equal to, or greater than, the time step.

- *Frequency of clearing* defines the frequency of market clearing, i.e. how often the market is cleared. Note that in order to take into account the most up-to-date system state data, market needs to be cleared sufficiently often. However, from an algorithmic perspective, market clearing frequency is constrained by how fast the clearing can be solved to achieve (near) optimum solution, with higher-frequency requirements (e.g. for security reasons) possibly yielding an economically suboptimal solution that can still be acceptable.

In the SmartNet simulator, time steps, time horizon and frequency of market clearing are parameters that are controlled by the user, providing necessary flexibility to adjust market operation and clearing for the particular conditions that will typically be dependent on regulatory settings. However, note that from 1 January 2025, the imbalance settlement period should be 15 min in all control areas. Since market operators (MOs) on the day-ahead market (DAM) and intraday market (IDM) shall provide the opportunity to trade energy in time intervals which are at least as short as the imbalance settlement, energy will be traded in at least 15 min period from 2025.

It is also important to consider that to create a level playing field for some technologies, such as wind or PV, trading should be moved as close as possible to operation to reduce influence of forecasting error, as will be discussed below, in Sect. 7.5.12. The issue of stochastic behaviour of some technologies can also be addressed through application of rolling horizon in the market clearing algorithm; however this will increase computational time and also provide different clearing outcomes compared to design that does not include rolling planning.

7.5.2 Objective Function for Electricity Market Design

As discussed in Chap. 3, there are a number of ways to define an objective function in a market clearing algorithm. The most common economic objectives (which are utilised in other energy and ancillary markets) are minimisation of activation cost and maximisation of social welfare. The former aims to reduce the cost to the system operator which is responsible to solve grid problems using the market product, while the latter, preferred by the EU regulators, seeks to increase the welfare of both sellers and buyers of flexibility services.

The approach adopted in the SmartNet seeks to minimise the total cost of deploying the flexibility (procurement and activation), and therefore, the objective function is defined as maximising the welfare by avoiding unnecessary activation. This means that only activations that contribute towards releasing congestion or voltage violations will be used, while avoiding bids that may seek to introduce a need for flexibility which will be solved by activating another flexibility bid.

7.5.3 Accounting for Technical DER Constraints in a Market Design

When deciding on modelling technical capabilities and responses from different DERs, it is important to decide which constrains should be included in the market model and how market participants should account for technical constraints of different DER technologies. One approach is to allow for the market design and optimisation formulation to directly account for these constrains, while the other is to expect market participants, and in particular aggregators, to develop bidding strategies that include those constraints indirectly. For example, the question is how to model ramping, binary states of some devices or, in the case of ICT, duration, size, response latency and activation time. Detailed discussion on aggregation models applied in the SmartNet project has been reviewed in Chap. 3 and [33].

Modelling of technical constraints of DREs also influences the definition of bids, i.e. products, used by market participants, and in particular aggregators, as will be discussed in Sect. 7.5.11.

7.5.4 Management of Voltage Constraints

In addition to technical constraints of DERs, it is also important to include limitations of the power network into the market model. This includes voltage control which is formally defined as non-frequency ancillary service [8], and thus shall be allowed to be procured by DSOs in market-based manner (both active and reactive powers can be used for voltage control).

Whereas DC approximation which neglects voltage issues can be used for transmission network operation, this type of approximation does not give accurate results for the distribution networks, where full AC network flow models are typically used. Therefore, inclusion of realistic physical models of the distribution system networks into a market clearing algorithm demands new approaches to satisfy computational tractability of the used algorithms.

Including power system's physics into a market clearing methodology implies solving an optimal power flow (OPF) problem, and in the case of distribution network, AC OPF, which is a complex non-linear computationally challenging task, especially in the presence of binary variables. To enable utilisation of existing solvers and provide computational tractability, modelling of the distribution networks requires new numerical analysis approaches. In the SmartNet, the simulator is based on DistFlow model, as discussed in detail in Chap. 3 and [34].

7.5.5 Pricing Mechanisms

In general, there are two main pricing methods that are used in the centralised electricity markets:

- The *pay-as-bid* approach where the activated bids receive the price corresponding to the activated quantity in the bidding curve
- The *pay-as-cleared* or marginal pricing approach where the activated bids receive the same price per MWh (or MW over a time step), corresponding to last activated/most expensive flexibility

However, as the considered power system is not a perfect copperplate, network constraints, both at the transmission and distribution levels, have to be taken into account. There are two approaches to how marginal pricing can reflect those network constraints:

- A *nodal* approach where a price for flexibility is associated with the most granular level in our network representation, i.e. to each node of the distribution grid
- A *zonal* approach where a price for flexibility is associated with a zone covering different nodes. Each zone can have a different price but the nodes in the same zone have the same price. There are no constrains within one zone, just between different zones.

Although, in general, market participants can submit any bid, and not just the true costs, economists suggest that under perfect market conditions the best strategy is for them to submit true costs as it will increase the chance of being selected, while the difference between the clearing price and true costs would yield a profit. However, bidding of market participants is not under control or regulated (unless under special circumstances, such as exercising market power), which means that they can submit bids that deviate from true costs. Under pay-as-bid pricing, bids have to include a profit margin, as payments and revenues are equal to the submitted bids.

In the SmartNet market model, the price received per activated bids is determined through the market clearing, which yields locational marginal prices (LMPs). In addition, note that in the SmartNet market model [35], there is a price for both upward and downward flexibilities for each node, both at transmission and distribution networks.

7.5.6 Relationship with Previous Markets and Impact of Gate Closure

It is important to define the related markets, the timing and effects they have on each other, as well as how the price signal from one market can affect the participation and result in the subsequent one. Markets which precede activations of ancillary services trades include:

- *Day-ahead (or spot) market* is represented by the day-ahead auction which typically closes up to 24 h before the operation. Since the forecast (of both production and consumption) begins to firm up, each player must place the expected exchange profile and can also include complex constraints. These constraints depend not only on the technology, but also on how a market design accounts for technical constraints of DERs (as discussed above in Sect. 7.5.2). The matching of traded consumption and production determines a price for each time block corresponding to the baseline price for electricity in the given time period. According to the clearing results, the same participants can then develop their trading strategies for the subsequent markets according to their expectations on price evolution and possible imbalance.
- *Intraday market* is aimed at providing participants with an opportunity to rebalance their position in case of deviations with respect to day-ahead profiles (in order to reduce their imbalance exposure). Intraday markets play an important role since forecasted profiles and grid circumstances that may affect market clearing become more accurate by approaching the real-time operation. These markets open shortly after the day-ahead auction closes and trades can then be made until the market closes shortly before delivery (gate closure). The latency between intraday gate closure and beginning of delivery is called "intraday lead time", and it is important to understand that it sets the time horizon in which no further intraday market actions are possible. In this time interval, unforeseen events and any deviation from the intraday market outcome are considered an "imbalance", unless it results from an activation in the next market which is for ancillary services.

There is widespread agreement that intraday markets should bring gate closure as close as possible to real time in order to allow, in particular but not only, RES to recalculate their actual generation taking into account the most updated forecasts [36]. This will reduce imbalances, with an important cost reduction by the point of view of the system and of producers as well, since they are called to pay for the activations needed to solve the imbalances they caused.

The gate closure of the other markets may have an impact on the ancillary services market. If they are cleared too close to real time, TSOs and DSOs may not be able to evaluate the status of the system and calculate the needed reserve in order to operate the AS market. To avoid this situation, strong investments are needed both in ICT, in order to allow detailed control on DN, and in computational power, in order to speed up the clearing of the market.

Furthermore, activation and ramp constraints may prevent some technologies from providing their services. In these cases, even if it could be expected that the technological improvement will remove those technical constraints for a large part of resources, for some others this is impossible (e.g. loads related to industrial processes). So an intervention from the regulator may be needed, in order to find out the chronological synchronisation of the different markets best suiting the needs of the whole power system.

Finally, we have to remark that implementing market schemes such as CS-E where flexibility resources are made available not only to system operators, but also to commercial market parties to solve their own flexibility problems, could be an effective way to tackle this issue, since they are a mix of intraday market and ASM.

This type of CSs has not been assessed in details within the SmartNet project due to the modelling complexity, and further investigation will be needed before making a decision for its implementation.

7.5.7 Local Congestion Management by DSO Versus Centralised TSO Market

One of the major questions related to TSO-DSO interactions is related to how to manage local congestion at the distribution level that may often be caused by DERs and their participation in provision of AS. The coordination schemes investigated in the SmartNet project, and described in detail in the previous chapters, look at both centralised and decentralised types of architectures.

In the centralised approach, it is the TSO who takes the activation decisions for flexibilities offered, not only at the transmission network level but also at the level of the distribution networks. In this case there is no DSO managed local market. Under the centralised model, if the TSO has full observability of the DERs and distribution network, it can include their constraints when clearing the global balancing market. Conversely, when there is no observability of distribution networks, TSO will clear the market without including such constraints, and DSOs will then need to approve or block the ancillary service during the counter-trading phase. The market clearing will not be run again, but rather the counter-trading is handled outside the market, with a limited time allowed for vetoes. In case of a veto, the TSO (or the DSO) can use its own resources to compensate for the assets which are not activated.

If a decentralised market architecture is used, and if the local market operators (DSOs) have to transfer the (additional) flexibility, offered by DERs or by aggregators, from the local market to the central AS market, a methodology is needed on how the DSO would perform this task. Although analysis in the SmartNet assumed that DSO would be ready to carry out local congestion management via a market mechanism or by contracting services and not only using their own resources, from that perspective, it is necessary to recognise the importance of a number of issues related to liquidity of local markets, as discussed in the next section.

7.5.8 Illiquidity of Local Networks

As mentioned above, liquidity of the local market is one of the crucial issues that needs to be resolved in order to ensure its operation. However, in contrast to wholesales and large markets, this may be more difficult to achieve both due to a lack of available resources and network constraints. It will also be important to recognise whether the illiquidity is caused by a genuine lack of resources, in which case either infrastructure reinforcement or some other regulatory solutions may be proposed to improve availability of flexibility. In the case it is due to abuse of market power, a regulator may need to intervene, just as in the case when such problems occur in the wholesale market.

A number of issues are related to liquidity of local markets, including:

- A local market may create competition for flexibility resources. However, the scope and the size of the local market may dictate its liquidity.
- The timing of the sessions (local market) is a factor that impacts the liquidity of this market and the markets managed by the TSO.
- Distinction between congestion and balancing reserves and coordinated actions between TSOs and DSOs towards flexibility would help reduce competing for the resource and enhance liquidity in both markets.
- Liquidity may not be realised due to the lack of advance reservation of capacity to the real-time market.
- Minimum bid sizes may be too large and the bid structures too complex.

To overcome the issue that can arise in some instances when the lack of flexibility may preclude SmartNet simulator to clear the market, very expensive virtual slack-type distributed generators have been introduced when evaluating system operation under different CSs. Note that utilisation of these generators will only help analyse the network operation and possible issues, while in real networks such a solution would not be a possibility. Nevertheless, implementation of transition towards CSs and local markets will have to be preceded by detail planning studies where activation of virtual slack DGs is an important indicator of potential market issues. However, the issue of liquidity is not related to pricing mechanisms, as it can happen both under marginal, i.e. pay-as-clear, and pay-as-bid pricing methods.

Therefore, if only a small number of resources are reliable and/or available to the system operator, two important risks arise:

1. Those resources may have potential to exercise market power, that is, to affect the market price making, and thus increase their profit at the expense of the system and other market participants.
2. The SO may not be able to solve congestion in the network under its control by means of the market and thus be forced to activate unwanted measures that are more expensive, thus increasing the costs for the system.

It is clear that the smaller the system, the higher the risk of illiquidity, so that for those schemes that consider local markets (CS-B and CS-C), this may be a key issue. The problem may also occur in the schemes that present global markets, in particular when congestions result in the separation of a very small portion of the system from the rest of the network.

In the SmartNet project, this issue has not been investigated due to the mathematical complexity of its modelling, but regulation authorities should take this problem into account.

The increase in prices consequent of (1) and (2) should be significant signals for the investors that installing new flexibility resources could be profitable. However, if the solution of the illiquidity is left to market signals alone, it could result in the classic boom-and-bust cycles, since high prices will attract significant investments and connections of DERs, which can then lead to a sharp drop in prices and, thus, sudden drop in new investments and DER connections.

This situation is usually not tolerable for the society due to the continuous changes in prices; neither it is good for the system that sees its reliability unstable; nor it is for investors, since the high uncertainty in prices makes investment costs increase. So regulation should find out other kinds of solutions to increase the amount of flexibility available to the system.

One of the possible solutions, especially suitable for small DSOs, is the possibility for them to join up in a single local market sufficiently large to avoid scarcity of liquidity. This approach should be encouraged not only because a larger pool of resources would be available, thus increasing competition and reducing the chances of market failure, but also because it will increase economic efficiency, since many small local markets have higher ICT costs than a few local markets with a reasonable size.

Another possible action that will increase the available flexibility could be the introduction of adequate market products, tailored to the technical needs of distributed energy resources (DERs), in particular to demand resources.

7.5.9 Operation of Possible Local Market: Single DSO Versus Common Distribution Market Operator

A number of CSs in the SmartNet assume existence of the local markets where a number of players, mainly aggregators or actors assuming such a role, will offer to trade flexibility. As have been discussed in previous chapters, market clearing procedure allows the participants to submit their upward and/or downward bids, and the most economic ones, subject to network operation constraints, are selected to provide necessary ancillary services. It has been assumed that the size of the local market was sufficiently large to avoid market illiquidity. However, it is also important to recognise other aspects that need to be considered in the operation of the local markets. This includes operation of network areas with multiple DSO, which vary in size and resources availability. For example, in some countries, such as Norway, there are a very high number of small DSOs, while in others there are few large DSOs. In some cases, there could even be DSOs that are operating as sub-DSOs. For small DSOs, possible solutions could include defining and implementing procurement mechanism for their own needs that is cost-efficient, or pooling resources with other similar neighbouring DSOs.

Therefore, a local market should be established only where it makes more sense, with small DSOs (e.g. with 100k consumers or less) not forced to implement measures impacting their cost structure without sound reasons to do so. Small DSOs should be given the option to decide if they need a local market and who the operator of such a market should be.

In addition, DSOs should be allowed to implement pilots that aim at testing different market designs and pricing schemes for services required at distribution level.

7.5.10 Prequalification of Resources in Distribution Networks

In general, some prequalification and initial capacity allocation may be necessary in order to guarantee that adequate amount of flexibility is available in all relevant locations and times. Otherwise, there is a risk of market failures. SmartNet concepts take into account the constraints during market clearing. Thus, post qualification is not needed nor applied, as it is implied in market clearing procedure.

The process of prequalification has been divided in two separate processes: technical prequalification and system prequalification. A technical prequalification validates the technical requirements of a unit that wants to participate to the AS market. System prequalification is defined as an upfront process where the DSO validates the participation of DER to the flexibility market, under the condition that it does not violate local grid constraints. Therefore, for the qualification of resources in distribution networks, distribution constraints must be taken into account. The DSO should be responsible for this, and in some cases it already does. In addition, it is important to recognise that prequalification can have a significant impact on which DERs can participate in provision of particular services, and, therefore, this process has to be carefully designed not to eliminate valuable resources if they can provide these services.

With respect to ancillary services, distribution grid constraints can be taken into account by the TSO ex ante, at the time of the clearing, or ex post. In addition, their consideration could be static, when the constraints are considered only once, e.g. at the connection phase, or dynamic, when there is a continuous check of the state of the distribution grid.

An optimal consideration of distribution grid constraints requires an active participation of the DSO in the procurement and activation process of ancillary services being sourced (in whole or in part) by distribution connected units. The need to consider distribution constraints in a static or dynamic way would mostly depend on how the distribution grid is regarded, i.e. as a copperplate or as a grid with limited distribution capacity. If no fundamental congestions exist, then it might be "enough" to take a static approach. However, the selection of a static approach when the distribution grid has limited capacity may reduce the amount of flexibility that can be used. Therefore, in this situation it might be better to opt for the dynamic approach.

7.5.11 Market Products

Ensuring level playing field in the participation of DERs, especially industrial loads, to the tertiary market is a key feature. This will result in an increase of the amount of flexibility at the SOs disposal, and would also solve other issues such as illiquidity, as discussed in Sect. 7.5.8.

For example, consider the case of a firm which can choose alternatively between two different production lines with two different consumption profiles. Its "flexibility" consists of choosing between the two different lines. Once one of the two profiles is chosen, it cannot be changed and has to be followed for the entire production time. Another example is related to thermostatically controlled loads (TCL) [10], which have to fulfil a heating (or cooling) demand: they are able to reduce (increase) their consumption profile for a few time intervals, after which they have to bring the temperature back within the predefined limits by increasing (reducing) their total energy consumption. The resources conisdere here are in general smaller and less flexible than those connected at the transmission level.

To allow the participation of DERs in ancillary services markets, it will become necessary to consider developing bidding products that can reflect their technical characteristics and also enable them to price their offers adequately.

These complex bids will, however, increase the complexity of the mathematical representation of the markets, as well as costs of the computational power needed to clear the market.

7.5.12 Influence of Forecasting Error

It is commonly expected that in 2030 and beyond resources at distribution level will be mainly composed by RES generation (e.g. PV power plants, mini-hydro, etc.). The reliability of these resources is strongly dependent on accuracy of their forecast generation outputs, and therefore the forecasting error has an important impact on the power system operation, regardless of which of the coordination schemes is implemented. However, as discussed above, the impact of the forecasting error may cause higher economic inefficiency for some coordination schemes compared to others. For example, it has the strongest impact in CS-D apporach, when a global market considers constraints of both distribution and transmission networks.

The most intuitive way to reduce the influence of forecasting error would be to increase forecasting reliability by using improved forecasting methodologies and better computational techniques. In addition, the performances and accuracy of forecasting techniques may significantly vary for different generation technologies, and in particular technologies strongly influenced by local climate and weather factors (e.g. mini-hydro) may be more prone to forecasting errors.

The precision of the forecasting error is not, however, under regulators' control, but a possible regulation action that can decrease the impact of forecasting error is to set the gate closure of ancillary services markets as close as possible to the real-time operation. Unfortunately, this can be done only up to a certain extent because some "dead" time is unavoidable. For instance, the ramp-up/ramp-down constraints of some resources may require allowing some addtional time for thier adjustments before the real time operation, as well as increased computational time needed to clear the market (which can be particularly relevant for schemes considering all of the network constraints).

7.6 Conclusions

Increased levels of distributed energy resources (DERs) and their participation in provision of ancillary services (AS) at both transmission and distribution levels call for a more advanced dispatching management of distribution networks to transform distribution from a "passive" into an "active" system. Moreover, new market architectures must be developed to enable participation of DERs in energy and ancillary services markets. New operational and trading arrangements will also affect the interface between transmission and distribution networks, which will have to be managed in a coordinated manner between TSOs and DSOs in order to ensure the highest efficiency, effectiveness and security.

Proposed TSO-DSO coordination schemes have been evaluated, tested and validated using a simulator that reflects market designs necessary to realise these schemes. In addition, pilot project in Italy, Denmark and Spain each tested one of the proposed schemes. Results indicate that the proposed schemes can enable realisation of low-carbon energy systems, and that the recently proposed regulatory frameworks are a move in a right direction. Besides evaluation of each of the schemes against the current regulation, it also needs to be recognised that there are a number of issues common for most of the proposed schemes, such as market design, including how to model technical and system operation constraints while still using tractable algorithms for market clearing. This also leads to a question of a design of bidding products that are suitable for different technologies, as well as how to address issues related to possible market illiquidity and market power. Also, there is a need to set up gate closer as close as possible to real-time operation to reduce forecasting errors which may cause inefficient market solutions.

It should be noted that the proposed schemes looked at different ways to realise TSO-DSO coordination, but the practical solutions adopted and implanted will depend on national system configuration, organisation of trading and system and network operators, as well as other particular needs that each national regulator will need to address.

References

1. D. Roberts, Network management systems for active distribution networks – A feasibility study. DTI Distributed Generation Programme (Contractor: SP PowerSystems LTD), Contract Number: K/EL/00310/00/00, URN Number: 04/1361 (2004)
2. European Commission, Clean energy for all Europeans (2016), https://ec.europa.eu/energy/sites/ener/files/documents/cleanenergy_com_en.pdf
3. European Commission, Clean energy for all Europeans – additional set of documents (2016), https://ec.europa.eu/energy/en/news/commission-proposes-new-rules-consumer-centred-clean-energy-transition
4. CEDC, E.DSO, ENTSO-E, Eurelectric, GEODE, An integrated approach to active system management: With focus on TSO-DSO coordination in congestion management and balancing. TSO-DSO report (Apr 2019)
5. European Parliament and Council of the European Union, Directive 96/92/EC concerning common rules for the internal market in electricity, Brussels, 19 Dec 1996
6. Directive 2003/54/EC of the European Parliament and the Council of 26 June 2003 concerning common rules for the internal market in electricity and repealing Directive 96/92/EC
7. Regulation (EC) No 714/2009 of the European Parliament and of the Council of 13 July 2009 on conditions for access to the network for cross-border exchanges in electricity and repealing Regulation (EC) No 1228/2003
8. Regulation (EC) No 713/2009 of the European Parliament and of the Council of 13 July 2009 establishing an Agency for the Cooperation of Energy Regulators
9. Directive (EC) 2009/72/EC of the European Parliament and of the Council of 13 July 2009 concerning common rules for the internal market in electricity and repealing Directive 2003/54/EC
10. Directive (EC) 2009/73/EC of the European Parliament and of the Council of 13 July 2009 concerning common rules for the internal market in natural gas and repealing Directive 2003/55/EC
11. Regulation (EC) No 715/2009 of the European Parliament and of the Council of 13 July 2009 on conditions for access to the natural gas transmission networks and repealing Regulation (EC) No 1775/2005
12. European Court of Auditors, Improving the security of energy supply by developing the internal energy market: more efforts needed. Special report no 16/2015, Luxembourg (2015), https://doi.org/10.2865/317301
13. EU Emissions Trading Scheme – documents available at: https://ec.europa.eu/clima/policies/ets_en
14. Communication from the Commission to the European Parliament, The Council, The European Economic and Social Committee, The Committee of the Regions and The European Investment Bank, A framework strategy for a resilient energy union with a forward-looking climate change policy, COM/2015/080 final (2015)
15. S. Horsmanheimo et al., D3.2 – ICT architecture design specification. Technical report, SmartNet (2017)
16. CEDEC, EDSO, ENTSOe, Eurelectric, Geode, General guidelines for reinforcing the cooperation between TSOs and DSOs (2015). [Online]. Available: https://docstore.entsoe.eu/Docments/Publications/Position%20papers%20and%20reports/entsoe_pp_TSO-DSO_web.pdf
17. ENTSO-E, Distributed flexibility and the value of TSO/DSO cooperation (2017). Available: https://www.entsoe.eu/Documents/Publications/Position%20papers%20and%20reports/entsoe_pp_DF_1712_web.pdf?Web=1
18. WindEurope, Views on the TSO-DSO coordination – Enabling flexibility from distributed wind power (2017), https://windeurope.org/wp-content/uploads/files/policy/position-papers/WindEurope-views-on-TSO-DSO-coordination.pdf

19. CEER, Position paper on the future DSO and TSO relationship (2016). Available: https://www.ceer.eu/documents/104400/-/-/e8532c60-56d5-17bf-cecb-b4f9594fe0c4
20. S.Y. Hadush, L. Meeus, DSO-TSO cooperation issues and solutions for distribution grid congestion management. RSCAS 2017/65, Robert Schuman Centre for Advanced Studies Florence School of Regulation (2017). Available: http://cadmus.eui.eu/bitstream/handle/1814/49744/RSCAS_2017_65.pdf?sequence=1&isAllowed=y
21. ETIP-SNET, Vision 2050 – Integrating smart networks for the energy transition: Serving society and protecting the environment (2018). [Online]. Available: https://www.etip-snet.eu/wp-content/uploads/2018/05/VISION2050-v10PTL.pdf
22. Final 10-year ETIP SNET R&I roadmap covering 2017-26 (2016), https://www.etip-snet.eu/wp-content/uploads/2017/03/Final_10_Year_ETIP-SNET_RI_Roadmap.pdf
23. A. Morch et al., D6.2: Evaluation on project results related to a number of models and roadmaps. Technical report, SmartNet project (2019)
24. I. Kockar et al., D6.1: Conclusions from national tests/simulations and their evaluations. Technical report, SmartNet project (2019)
25. D. Safice et al., D6.3: Policy recommendations to implement and/or overcome barriers and enable TSO/DSO integration, which, which will conclude the work package and elaborate the final guidelines and regulatory recommendations. Technical report, SmartNet project (2019)
26. H. Gerard, E. Rivero, D. Six, D1.3 – Basic schemes for TSO-DSO coordination and ancillary services provision. Technical report, SmartNet project (2016)
27. RSE, SelNet, Selta, Technalia, TERNA, D5.1 – Results of Pilot A (Italy). Technical report, SmartNet project (2019)
28. A.G. Azar et al., D5.2 – Results of Pilot B (Denmark). Technical report, SmartNet (2019)
29. M. Pardo, M. Duarte, P. Paradell, C. Madina, J. Jimeno, M. Marroquin, A. Ibànez, E. Estrade, L. Jones, D5.3 – Results of Pilot C (Spain). Technical report, SmartNet (2019)
30. AIT, N-Side, RSE, Technalia, VITO, VTT, D4.3 – Cost-benefit analysis of the selected national cases. Technical report, SmartNet (2019)
31. F. Pröstl Andrén et al., D4.4 – Lab-environment set-up and simulations. Technical report, SmartNet (2019)
32. European Commission, Proposal for a Directive of the European Parliament and of the Council on common rules for the internal market in electricity (23 Feb 2017). [Online]. Available: https://eur-lex.europa.eu/legal-content/EN/TXT/?uri=CELEX%3A52016PC0864R%2801%29
33. M. Dzamarija et al., D2.1 – Aggregation models. Technical report, SmartNet (2018)
34. G. Leclercq et al., D2.2 – Network and market models. Technical report, SmartNet (2019)
35. M. Rossi et al., D4.2 – Scenario set up and simulations results. Technical report, SmartNet (2019)
36. A. Morch et al., Post-2020 framework for a liberalised electricity market with a large share of renewable energy sources. Final Publication from Market4RES project (Oct 2016), http://market4res.eu/wp-content/uploads/LR-Market4RES-final-publication.pdf

Chapter 8
Conclusions

Gianluigi Migliavacca and Ilaria Conti

In conclusion of the long path on TSO-DSO interaction and ancillary services procurement covered by the previous chapter of the book, it could be interesting to re-propose the set of questions to which the Horizon2020 project SmartNet aims to give an insight:

- Which ancillary services could be provided from entities located in distribution networks?
- Which optimised modalities should be used for managing the network at the TSO-DSO interface?
- How should the architectures of dispatching service markets be consequently revised?
- What ICT infrastructure installed at the distribution-transmission border can guarantee observability and control?
- Which could be the regulatory implications?

Now, it is time to focus on the last question: What are the regulatory implications of procuring ancillary services from distribution networks? The starting point is the synthetic view that can be gained through the 3.5 years of activities of the SmartNet project, which can be condensed in the following statements:

- *Traditional TSO-centric schemes could stay optimal if distribution networks don't show significant congestion* not unlikely in near-future scenarios, since distribution grid planning was (and still is) affected by the fit-and-forget reinforcement policy. In a first period, costs to implement monitoring and control

G. Migliavacca (✉)
Energy System Development Department, Ricerca sul Sistema Energetico, Milano, Italy
e-mail: Gianluigi.Migliavacca@rse-web.it

I. Conti
FSR, Florence, Italy

G. Migliavacca (ed.), *TSO-DSO Interactions and Ancillary Services in Electricity Transmission and Distribution Networks*,
https://doi.org/10.1007/978-3-030-29203-4_8

systems within distribution networks could result higher than the effect of over-investment inefficiencies due to the old fit-and-forget philosophy. This could engender resistance in some DSOs to consider flexibility as a value. This could also call for a revision of present remuneration schemes for DSOs' investments, so that they can claim OPEX and not only CAPEX.

- *More advanced centralised schemes incorporating distribution constraints show higher economic performances, but their performance could be undermined by big forecasting errors*, which could bring them to take wrong decisions. As distributed generation, constituting a good share of the possible service providers in distribution, is mainly composed by RES generation (e.g. PV power plants, mini-hydro, etc.), it is important that the gate closure is shifted as much as possible towards real time and forecasting techniques are improved. Such techniques can be better for some generation technologies (PV) but much worse for others which are strongly influenced by local factors (mini-hydro).
- *Technical reasons and high ICT costs dis-advise to give balancing responsibility to DSOs*. Nonetheless, the sheer economic performance of such shared responsibility schemes is not always bad (sometimes separating transmission and distribution markets could prevent high prices in one area to be spread to the other).
- *Decentralised schemes are usually less efficient than centralised ones* because the two-step process introduces undue rigidities. Scarcity of liquidity and potential impact of local market power, along with extra constraints introduced to avoid counteracting actions between local congestion market and balancing market (e.g. increasing system imbalance while solving local congestion), furthermore negatively affect economic efficiency of decentralised schemes.
- *Decentralised schemes request to put in place further coordination actions between TSO and DSO*: resources which are bid in both sequenced markets should not be selected twice (a "common marketplace" mechanism should be implemented).
- *Local congestion markets should have a "reasonable" size and guarantee a sufficient number of actors are in competition* in order to prevent scarcity of liquidity and exercise of local market power. For that, *small DSOs should pool up* in order to create a common congestion management market: too many small local markets would increase ICT costs and reduce competition, with detrimental effects.
- *Intraday markets should bring gate closure as close as possible to real time. However, it is not feasible to overlap a real-time session of intraday market with a service market*: this solution would create uncertainty in the operators (TSO and DSO) in charge of purchasing network services because they would be no longer sure of how many resources are needed (i.e. the real amount of congestion and imbalance). For this reason this coordination scheme is strongly dis-advised.
- *Balancing and congestion markets should have as target not to optimise system social welfare (i.e. by contrast, the goal of energy markets) but just to buy the minimum amount of resources to get the needed network services while perturbing the least possible the results of the energy markets*. This advises against

allowing the award of sets of balanced upward and downward bids just to reduce total costs ("market arbitrage") even whenever this could reduce total system costs.

- *Ensuring level playing field in the participation of distributed resources (especially industrial loads) to the tertiary market means to be able to incorporate into the market products some peculiarities of such resources* (loads or generators) without which it is nearly impossible for them to participate. This could imply to enable complex bids or other sophisticated products.
- *Reaction to commands coming from TSO or DSO in real time of the control loops which were initially planned for real-time services provision can be too slow*. So, a testing is needed to ensure compatibility with requested reaction times.
- *ICT is nearly never an issue: whatsoever TSO-DSO coordination scheme is implemented*, the economic performance depends by wide and large on operational costs. For all coordination schemes, ICT costs stay one order of magnitude lower than operational costs.

So, from the previous bullets, it can be argued that the TSO-DSO coordination schemes featuring a local market could present criticalities dis-advising their adoption (liquidity issues, potential for exercise of market power from incumbent service providers, dis-optimality in the economical dispatch deriving from the market clearing). The adoption of a centralised "complete" scheme including distribution transit constraints in the real-time market architectures (balancing and congestion management) can be for sure a very good solution provided that congestion is not a rare case in distribution networks; otherwise the "status-quo" centralised model where real-time markets are managed by the TSO and don't include distribution networks constraints could be equivalent (or even perform slightly better in case of important forecasting errors). So, in order to enable a good functioning of the "complete" model, two important preliminary facts should materialise:

- Gate closure of energy markets (day ahead and intraday) should be as close as possible to real time in order to reduce to the minimum the incidence of forecasting errors.
- Distribution networks planning should abandon the fit-and-forget policy bringing to oversizing the network to prevent congestion, and a more cautious investment policy should be put in place, where a trade-off assessment between maintaining congestion and managing it in real time and prevent it by new investments should be systematically applied.
- Real-time monitoring and control devices should systematically be deployed in distribution networks, allowing the system to know in real time their status. This will require important investments in the next years.

Finally, yet materialising the right conditions for profitably applying the "complete" TSO-DSO coordination model, its "depth" (so the lowest voltage level which can be seen in detail with all nodes and all transit constraints completely represented in the real-time market clearing) can't but be limited: we can reach medium voltage but for sure not the lowest voltage levels characteristic of the domestic loads. For the lowest voltage levels which cannot be reached in detail but are seen by the market as a "bus bar" system, the approach suggested by the Italian pilot (see Chap. 6) and consisting in the evaluation of the capability curve, in order to allow the TSO to know the available active and reactive power margin on DSO network considering the capability of each power plant and the operational limits of the distribution grid, can be the optimal one.

8.1 TSO-DSO Interaction: The EU Regulatory Framework of Reference

Now, turning back to our introductory question – what are the regulatory implications of procuring ancillary services from distribution networks? – it is essential to make one step back and look at the regulatory framework we're making our steps in.

As we will see in a moment, the issue of electricity TSO-DSO interaction has been the focus of great regulatory attention in recent years, and the debate is now being extended (or "mirrored") to the gas sector.

The most recent EU regulatory provisions concerning the interaction between TSOs and DSOs are included in the European Commission's Clean Energy Package (CEP) [2][3], which was adopted in May 2019 and just entered into force.[1]

The CEP includes clear provisions that will enable DSOs to procure flexibility services ([4])[2,3] and aims to define the conditions under which they may acquire them, without distorting the market for such services [5].

In this sense, the new package introduces an innovative approach since, in many European countries, there are currently no rules in place that allow DSOs to procure flexibility services.

Before the Clean Energy Package entered into force, the most recent regulations of reference on these issues were (and still are) the electricity network codes (NCs), namely, the capacity allocation and congestion management (CACM), forward capacity allocation (FCA), electricity system balancing guideline (EBGL) and system operation guideline (SOGL). The provisions included in these NCs aim at fostering cooperation among TSOs, with the scope to unlock flexibility resources connected to any part of the European power grid and efficiently manage grid constraints.

Notably, the system operation guideline (for electricity), in addition to several provisions requiring coordination between TSOs and DSOs, includes a general obli-

[1] With a date of application of 1 January 2020 for the electricity regulation, while the electricity directive will have to be transposed into national law within 18 months by EU member states.

[2] By "flexibility services," we intend non-frequency ones and congestion management services. The CEP excludes indeed the possibility for DSOs to procure frequency ancillary services (frequency containment reserves (FCR), frequency restoration reserves (FRR) and replacement reserves (RR)).

[3] Proposal for a Regulation of the European Parliament and of the Council on the Internal Market for Electricity. Vol. 379. https://ec.europa.eu/energy/sites/ener/files/documents/1_en_act_part1_v9.pdf.

gation for TSOs to consult DSOs and take into account the potential impact on their system – concerning, in particular, the following areas:

- Data exchange: obligation for TSOs and DSOs to agree on the scope, processes, formats, etc.
- Coordination in preparation and activation of remedial actions
- Coordination in pre-qualification and activation of reserves from units connected to the DSO grid
- Exchange of information related to infrastructure projects and coordination planning for outages [1]

8.2 The Novelties Introduced with the Clean Energy Package

As previously mentioned, the CEP introduces the possibility – for the first time in several EU member states – for DSOs to procure flexibility services.

This provision is expressed in a very precise and non-contestable manner in article 32.1: "Member States shall provide the necessary regulatory framework to allow and incentivise distribution system operators to procure services in order to improve efficiencies in the operation and development of the distribution system, including local congestion management".

Moreover, flexibility services shall be procured by DSOs with a *transparent, non-discriminatory and market-based procedure*.

The *transparency* requirement implies the need for member states to "*define* the exact Regulatory framework" (Art 32(1) of the Electricity Directive) allowing and incentivising DSOs to provide flexibility services, including congestion management, in an efficient and sustainable manner.

The concept of *market-based flexibility*, according to Nouicer and Meeus [7], is a process whereby flexibility is obtained and priced through an independent market mechanism from all stakeholders that are a source of flexibility, benefit from it, or have a controlling role, i.e. consumers, producers, BRP, system operators and regulators.

Finally, the non-discriminatory requirement is based on the need to include as potential sources "all market participants including renewable energy sources, demand response, energy storage facilities and market participants engaged in aggregation" as stated in *Art 32(1a)*.

8.3 Coordination with TSOs in the Procurement of Flexibility Services

Regarding coordination between TSOs and DSOs, the provisions included in the Clean Energy Package mainly focus on avoiding overlaps, by ensuring an efficient data exchange on the available flexibility resources and hence avoiding a double activation from a DSO and a TSO of the same flexibility source.

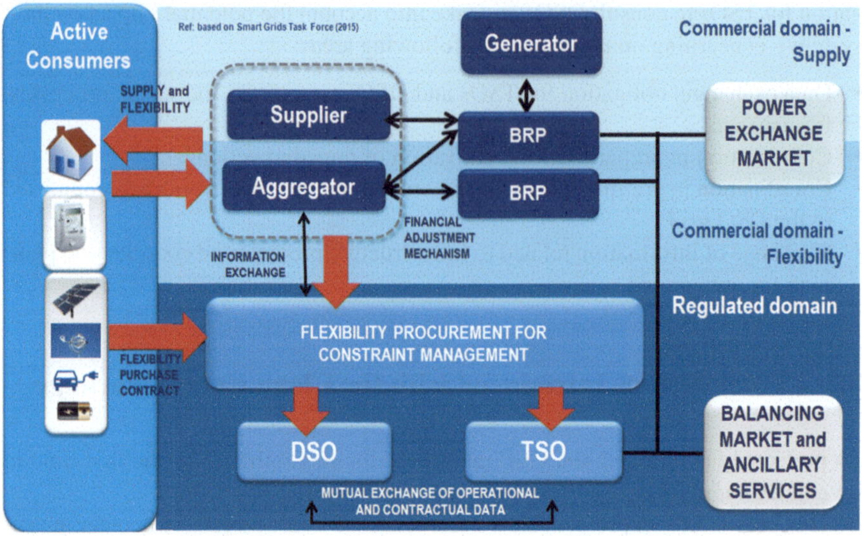

Fig. 8.1 A one-system approach for flexibility procurement. (Source: [6])

According to *Art 32(1)* of the E-Directive, "distribution system operators shall exchange all necessary information and coordinate with transmission system operators in order to ensure the optimal utilisation of resources, ensure the secure and efficient operation of the system and facilitate market development".

In addition, for the access to flexibility resources, *Art 53(2)* of the E-Regulation states that "transmission and distribution system operators shall cooperate with each other in order to achieve coordinated access to resources such as distributed generation, energy storage or demand response that may support particular needs of both the distribution system and the transmission system". Figure 8.1 shows how flexibility could be provided in a one-system approach [7].

8.4 Towards the Future: TSO-DSO Interaction in a Multi-Energy System

It is striking to realise that when the SmartNet project started, in January 2016, the Clean Energy Package – as we just illustrated, the only existing regulation addressing the interaction between TSOs and DSOs – had not even been proposed by the EU Commission at that time. Its publication in November 2016, indeed, caught great attention and raised interest among the energy stakeholders because of the change of approach to the traditional regulatory setting and of its forward-looking perspective.

But the energy world has started changing so rapidly and so dramatically that – although the CEP was conceived only in 2016 – in these 3 years the regulatory debate has already evolved and evolved around an even more innovative concept, totally disruptive against the past: sector coupling.

Although no official or universally recognised definition yet exists, by "sector coupling", we intend the closer and closer integration of the electricity and gas sectors in the future EU energy system – which would ideally tend towards the integration also with other sectors or services (telecommunications, transport, heating etc.).

This profound transformation of the EU electricity and gas traditional structures into a multi-carrier energy system would clearly entail a substantial "upgrading" of the EU energy infrastructure, governance and regulation currently in place.

The concept of sector coupling, launched and strongly supported by the European Commission, originates from the need to meet the climate targets and Paris Agreement commitment: in order to achieve a fully decarbonised EU gas market, a strong interaction with the electricity system will make the transition smoother, faster and more efficient.

The several questions on how this intuitive concept will translate into concrete policy and regulatory measures is intensely discussed in a number of fora, conferences and workshops and will probably receive some partial answers already with the upcoming Gas Package 2020.

While it's too early to make an attempt at forecasting the detailed content of the package and the issues that it will aim to address, some extended considerations can probably already be done regarding concerning some of the points we analysed in the context of the SmartNet project.

8.5 TSO-DSO Interaction in the Gas Sector, Current Provisions

In the gas system, the interaction between TSOs and DSOs is not that different – in regulatory terms – from that taking place on the electricity side. System operation in gas is understood to be about securing a reliable flow of gas through networks to customers.

European countries differ in terms of size and number of TSOs and DSOs in a country, the level of unbundling and competition in the retail market and a number of other elements [1].

Regulation-wise, gas DSOs are obliged to regularly communicate with TSOs regarding information on intraday and daily metered inputs and off-takes in the distribution system. DSOs and TSOs must cooperate to provide network users with forecast, near real-time and allocation data on their gas portfolios. This provision is key as it allows network users to take responsibility to balance their portfolios, hence minimising the network operators' intervention, as foreseen by the Gas Balancing Network Code.[4]

[4] BAL NC.

Just like for electricity, other NCs (interoperability and data exchange) regulate more practical aspects of the communication flow between TSOs and DSOs (data format and data exchange) and also the physical attributes of the gas itself (e.g. pressure, odorisation).

What is substantially different with electricity, of course, is time.

While in electricity the information on network status needs to be constant, as balancing happens on a second by second basis, in gas everything moves much more slowly and networks can rely on storage (linepack) for balancing purposes.

8.6 Cross-Network Challenges and Solutions

Three of the main, certain, features that the future multi-energy system will have are:

- Greater importance of the local dimension of energy generation: this will be the case also for the gas sector, because of the increasing role played by renewable gases (biogas, biomethane, hydrogen, etc.)
- The increased need for an efficient and complex exchange of data (in digital format) at cross-sectoral level
- At energy transportation level, a greater need for interaction between TSOs and DSOs

The SmartNet finding that "decentralised schemes request to put in place further coordination actions between TSO and DSO" can therefore be extended to the context of a multi-energy system.

As CEER already envisaged 3 years ago, "As well as sharing information on operational data and network status, it is critical that DSOs and TSOs develop a constructive ongoing dialogue, which enables them to understand which system operation actions could have cross network impacts" [1].

On the other hand, as we're heading towards a more and more decentralised energy system, it will be important to clarify roles and responsibilities of the various market actors. As we noted during the SmartNet project, "decentralised schemes are usually less efficient than centralised ones", so it will be essential to compensate the decreased efficiency with greater efforts in optimising the information exchange.

> There will also be a range of actions that could offer cross-network solutions. For instance, a DSO may be able to reconfigure their network in a particular area to ease a TSO constraint. DSOs and TSOs should build a common understanding of (i) which actions undertaken by one party could have an impact on the other; and (ii) which actions of one party could support the needs of the other. [1]

8.7 Coordinating Interaction to Maximise the Whole System Efficiency

To the ultimate benefit of customers, who should be enabled to take the best informed decisions, DSOs and TSOs should define processes when cross-system impacts are anticipated and subsequently take appropriate actions, including actions that the DSOs and TSOs may need to take to support system operation.

The role of the regulator is important in this context, in ensuring that the regulatory framework does not constitute an obstacle to the access to flexible resources across the system.

Last but not least, an optimised interaction between TSOs and DSOs in the future energy system will be fundamental in case of emergency: as we have seen in SmartNet, with the increasing intermittence and unpredictability of RES sources, coordination becomes indispensable. The coupling with the gas sector at infrastructure level, in this sense, is promising as it may be able to provide the flexibility that electricity generation may need (coming from storage of renewable gases and, in a transitory phase, also from traditional gas sources).

References

1. CEER, Position paper on the future DSO and TSO relationship (2016)
2. European Commission, Proposal for a Directive of the European Parliament and of the Council on common rules for the internal market in electricity (recast) (2016)
3. European Commission, Clean energy for all Europeans: Commission welcomes European Parliament's adoption of new electricity market design proposals, Brussels 26 Mar 2019. Press Release (2019)
4. Florence School of Regulation, The changing role of DSOs in the Energy Union. Workshop material (2016), http://fsr.eui.eu/event/changing-role-dsos-energy-union/
5. S. Hadush, L. Meeus, DSO-TSO cooperation issues and solutions for distribution grid congestion management. EUI Working Paper RSCAS 2017/65 (2017), http://hdl.handle.net/1814/49744
6. L. Meeus, A. Nouicer, *The EU Clean Energy Package* (European University Institute, Florence, 2018). https://doi.org/10.2870/013463
7. A. Nouicer, L. Meeus, EU clean energy package: Course text for the Florence School of Regulation online course. Version of Apr 2019 (2019)

Index

G. Migliavacca (ed.), *TSO-DSO Interactions and Ancillary Services in Electricity Transmission and Distribution Networks*,
https://doi.org/10.1007/978-3-030-29203-4

The manufacturer's authorised representative in the EU is Springer Nature Customer Service Centre GmbH, Europaplatz 3, 69115 Heidelberg, Germany. If you have any concerns regarding our products, please contact ProductSafety@springernature.com

Printed and bound by CPI Group (UK) Ltd, Croydon, CR0 4YY

30/09/2026

02225657-0001